Choose Your Baby's Sex

CHOOSE
YOUR BABY'S SEX

The one sex-selection method that works

By David M. Rorvik
and Landrum B. Shettles, M.D., Ph.D.

Illustrated with photographs and charts

DODD, MEAD & COMPANY
NEW YORK

1   2   3   4   5   6   7   8   9   10

Library of Congress Cataloging in Publication Data

Rorvik, David M
    Choose your baby's sex.

    First ed. published in 1970 under title: Your baby's sex.
    Bibliography: p.
    Includes index.
    1. Sex—Cause and determination. 2. Sex
chromosomes. I. Shettles, Landrum Brewer, 1909-
joint author. II. Title. III. Title: Sex-selection
method that works.
QP251.R62 1976   613.9′4   76-25195
ISBN 0-396-07356-5

In 1970, we dedicated our first book, *Your Baby's Sex: Now You Can Choose,* "to the overthrow of the so-called '50–50 Club,' an invention of complacent baby doctors who erroneously tell their patients that as a 'gift' of Nature they have a 50 percent chance of begetting offspring of the desired sex, and that beyond this, nothing can be done for them."

We repeat that dedication, this time with the added pleasurable knowledge that many thousands of prospective parents have since broken out of the "club" and that a growing number of researchers have confirmed Dr. Shettles' conviction that almost *all* of us can do so, if we will expend a little time and effort.

Acknowledgments

Special thanks is herewith extended to the Educational Foundation of America, a nonprofit organization, for its generous financial support of Dr. Shettles' continuing effort to perfect ever more accurate sex-selection methods.

The foundation has contributed significantly to many important causes, including improved quality of education, cancer research, and population control.

Contents

x   **Contents**

Choose Your Baby's Sex

NOTE ON THE TEXT:

In any collaboration there is often, in the mind of the reader, confusion about which author is speaking at any given point. Throughout this book, it should be understood that Mr. Rorvik is doing the observing and writing, while Dr. Shettles is providing the raw material and the medical expertise. This arrangement enables Mr. Rorvik to practice his profession (writing) and Dr. Shettles to pursue his (research). It also enables Mr. Rorvik to thrust, defend, and attack where Dr. Shettles might be restrained by his modesty or sense of professional propriety to remain silent.

CHAPTER 1

Who Is Laughing Now?

The following letter was written to the publisher of this book by Mrs. David Uhlman of Youngstown, Ohio:

October 13, 1976

Dear Sirs:

About three years ago a relative of ours (who worked in gynecology and obstetrics) told us of research she had heard . . . and of certain methods for conceiving a boy or girl, which we wrote down and kept for future use. When we decided to start a family I also bought baby books and one in particular mentioned Dr. Shettles' theory. My husband and I followed his recommended steps for having a boy.

We were so excited we told everyone what we had heard from the relative and what we had read in a book. They only laughed at us. Nevertheless, we now have a beautiful 16 month old baby boy and are con-

vinced that Dr. Shettles' theory is true and it wasn't
just "luck" that we had a boy.

A million thanks to Dr. Shettles and Mr. Rorvik, whose discoveries, we're sure, will drastically reduce the problems of over-population in the world.
Their book will remain a treasure in our home for-
ever.

Individual testimonials like Mrs. Uhlman's cannot by
themselves prove or disprove a theory. But if there are a
great many of them and if they tend to agree with the scientific findings of several independent researchers, then
certainly they must be reckoned with.

Editors, publishers, writers, and market analysts know
from experience that if even one or two percent of the
readers of a book or article are happy or mad enough to
write a letter about it, the publication in question is
"pulling," having impact. The various authorities differ
over whether the happy readers or the angry ones are more
likely to write.

When our book, *Your Baby's Sex: Now You Can Choose,*
was published in April of 1970, we believed that most of
our mail would come from irate individuals who had tried
our sex-selection method and had failed. From clinical
trials, we believed that as many as 20 percent could fail.
But since we stated that at least 80 percent would succeed
and, given a really conscientious effort, 90 percent or more
could succeed, we had some reason to fear that *everybody*
would expect to succeed. Thus we expected to hear pri-

marily from those who hadn't—and they would be mad as hell. The successful ones, we guessed, would have less motivation to write.

But, in fact, since 1970 we have received nearly 3,000 letters and many phone calls from those who have tried the method. Fewer than a dozen have reported failure. We were—and remain—dumbfounded, not because we believe that there have been only twelve failures, but simply because we expected the responses, at a bare minimum, to run 20 percent negative. Without claiming near infallibility, we nonetheless believe that we did not give our readers enough credit, and that most of the time they were willing and able to follow our sometimes complicated instructions to the T, ensuring success more often than we had expected.

It will be of interest to many of our readers (old and new) and particularly to the sociologists concerned about the effect of sex-selection techniques on sex ratios (a bumper crop of boys and a shortage of girls is one common fear) that nearly 70 percent of those responding were parents of one child who were seeking a child of the other sex for "the perfect balance." And having obtained a second child of the desired sex, they said they were satisfied and would have no more children. Of the roughly 30 percent with two or more children, almost all had children of the same sex, and when they finally gave birth to a child of the opposite sex, most also stated that they would have no more children. On the basis of such findings and other material that we will present later, we believe this technique may become a potent tool for population control.

But back now to the experiences of our readers. New readers may find encouragement—and some idea of what is possible—in the following sampling of letters in response to our first book.

December 5, 1973

Dear Dr. Shettles:

This letter is to tell you of the birth November 14 of our fifth child and first son. . . .

We followed your suggestions and have the boy we wanted. It seemed important to let you know of his birth and to thank you for your help. Now we just have to figure out how to raise a boy!

Most gratefully,
Perry Vartanian
(*Address withheld by request*)

October 16, 1975

Dear Dr. Shettles:

Please include us in your list of successes; we got a child of the sex we desired. After having had a boy we wanted a girl, and Ellen was born earlier this year.

The diaphragm we had been using for birth control simplified things. We found the correct day of ovulation and ceased having intercourse two days prior to it. As prescribed, we used the acid douche, I had no orgasm and shallow penetration and the missionary position were employed.

Thanks so much for what you have done to help people like us.

Dr. and Mrs. J.L. Jarum
(Address withheld by request)

September 29, 1971

Dear Dr. Shettles:

Thank you! After many years of questions and four female children later, your theory has worked for our benefit. Your book came at a time when we had decided there was no hope of ever having a male child. Armed with your procedure on how to conceive a male child, we decided to try one last time. The addition of our new son has indeed changed our lives and has convinced us that you have found a solution to the problem of repetitious sex of children among parents looking for a balance in females and males.

Most appreciative,
William F. Menz, Jr.
Sharon, Pa.

May 1, 1971

Dear Dr. Shettles:

I am writing to thank you, on behalf of my husband and our son, for the newest member of our family: a girl.

We are very grateful because it was always our plan to have two children, one of each sex. This is the

ideal family and one that we can provide for finan-
cially.

A great many of our friends are trying your
methods and are at different stages of pregnancy. I
believe many of them will be as successful as we have
been with your methods.

Sincere thanks,
Mr. and Mrs. Arthur Ramsey
(*Address withheld by request*)

October 19, 1974

Dear Dr. Shettles:

I wish to thank you for sharing your scientific
and medical knowledge with the public. My husband
and I have a new little son in our family of four
daughters. We carefully carried out your directions
on how to have a boy.

Sincerely yours,
Carolyn Pappas
Arcadia, Calif.

June 12, 1973

Dear Dr. Shettles:

I just wanted to take this opportunity to tell you
of our experiences after reading your book. I am the
proud father of three girls. For my fourth and final
chance to have a little boy, my wife and I purchased
a copy of your book. We followed it intensively and
we had a beautiful healthy baby boy.

We loaned your book to one of our friends who is a prominent ophthalmologist. He happens to be the father of three boys. He also followed your book and *voilà!* he fathered a baby daughter. You are batting 1.000. Keep up the good work! Many thanks.

> Very truly yours,
> Murray A. Schneider
> *Clayton, Mo.*

May 2, 1972

Dear Dr. Shettles:

I read and studied your book and . . . conceived using your method of obtaining a female. Happy to say, on April 10 my healthy daughter was born—much to the surprise of everyone on my husband's side, since I have a two-year-old son, and no girls have ever been recorded on my husband's side for many generations. Thank you for all your help.

> Sincerely,
> Judith C. Duncan
> *Mineola, N.Y.*

May 28, 1972

Dear Dr. Shettles:

Dr. James C. Burt suggested that we write thanking you for your research. You see, due to your work, we now have Kathy. After three sons, we are thrilled to finally have a beautiful daughter. We could never fully express our appreciation.

Dr. Burt holds you in very high esteem. He always said you were a fine, dedicated researcher. We certainly agree. Thank you so much for working to see that couples have sons *and* daughters.

Mrs. Robert Moots
Dayton, Ohio

February 1, 1971

Dear Dr. Shettles:

Last April I read about your theories on determining the sex of a child. My husband and I were planning on conceiving a second child at about that time and since our first was a girl, we wanted this one to be a boy. We therefore decided we had nothing to lose by following the procedures you outlined for conceiving a male child. Beginning with my last period we abstained from intercourse and began keeping a temperature chart.

On the thirteenth day of my cycle my temperature dipped three-tenths of a degree. (It had closely followed the day-by-day variations you described.) On that day, after [the prescribed] douche, my husband and I had intercourse with both of us achieving orgasm. I became pregnant that month and gave birth to a son on January 19—healthy and weighing eight pounds, thirteen ounces.

Needless to say, we are delighted because we have completed the family we had planned. We are also very grateful to you and thought the least we could do would be to write you and let you add our son to whatever statistics you may be compiling in support of your theories.

> Very truly yours,
> Mrs. Frank A. Muller
> *Palos Heights, Ill.*

June 18, 1975

Dear Dr. Shettles:

Please add our names to your list of "fans." After our son was born in 1972, we decided we would have one more child—hopefully a girl. I began seeking information from friends. Finally, your book was recommended, and I read it in February of 1974. . . . About April or May we began following directions. . . . I can practically pinpoint when our beautiful daughter was conceived, and she was born May 1. . . . I have informed my doctors and friends of your book and ordered it for two libraries.

By the way, we never—in the times we tried for a girl—deviated from the douche instructions. Some couples may say, oh, too much trouble, etc., but we had a good sex relationship nonetheless and have a beautiful result! Thank you for your wonderful book, and keep up the good work.

> Sincerely,
> Dr. Karen Fanta Zumbrunn
> *Emerson, N.J.*

May 16, 1974

Dear Dr. Shettles:

Last July I telephoned to ask whether you had made any refinements in your sex selection theories since the publication of your book, and you were kind enough to speak with me at length. Having talked with you about the hostility of [some of] your professional colleagues to your discoveries, I thought you would like to know that at least you have two more satisfied customers, as we are now the parents of a daughter, born April 26. The conception of our son, born before the publication of your method, also corresponds exactly to your predictions. I am sure you noted with interest the newly released study indicating that if couples could choose the sex of their offspring there would not occur the imbalance of the sexes which some of your colleagues seem to fear.

Yours sincerely,
Lynn Hecht Schafran
New York, N.Y.

December 13, 1975

Dear Dr. Shettles:

Our first child was a boy. Naturally, we wanted our second to be a girl. We followed your procedures and now have a beautiful daughter.

I didn't tell our doctor about this until I was on the delivery table. When it turned out to be a girl he was almost as excited as I was. Some of the other skeptical doctors have become a little more interested, too. We now consider our family complete and perfectly balanced. Thank you.

Jane Lyon
(Address withheld by request)

January 22, 1976

Dear Mr. Rorvik:

Last Thursday I called Dr. Shettles to inform him of the birth of my second son on January 14, and he suggested that I write you a note and give you the information that my wife and I used his method of abstinence, baking-soda douche, and intercourse all on the day of ovulation in order to increase our chances of having a son.

After we had three daughters in a row, my wife became very depressed, for she had wanted a son very badly. She began inquiring and asking various physician friends specializing in obstetrics and gynecology, and no one gave us any encouragement. She then talked with a researcher at the University of Michigan who informed her that Dr. Shettles was doing some work on selecting sex and it was about that time she was sent a copy of *New York* magazine containing an article written about Dr. Shettles. Shortly thereafter we also read the article written by Dr. Shettles and yourself in *Look* magazine.

In 1970 we took a trip to New York and met with Dr. Shettles and shortly thereafter we began using his method. Our first son, Jason, was born in March of 1971. Then last week, January 14, our second son was born.

In looking back and reviewing in our minds the conception of our three daughters, it was easy to understand how the chances of their being girls at conception were greater. I hope this information is of some value to you as I understand you are currently preparing a new edition of the book.

Sincerely,

S. William Paris, M.D.

Farmington, Michigan

And representative of the dozen "failures" we heard from:

September 8, 1975

Dear Mr. Rorvik:

My husband and I have used the method of sex selection described in your book, written with Dr. Landrum Shettles. We carefully followed the prescribed procedure for the desired sex, but it did not work for us.

We want to try again, using Dr. Shettles' technique, but before doing so I would be grateful if you

could advise me of any further findings in addition
to your 1970 book, which might help us.

Yours truly,

(*Name withheld by request*)

The writer went on to ask specific questions about timing ovulation, which we hope are answered in sufficient
detail in this book.

Whatever "opposition" has surfaced has come not from
readers but from a few doctors who, in our judgment, have
made negative pronouncements on the basis of misinformation or no information whatsoever. Nurturing precon-
ceived notions (such as, "There's nothing we can do about
the sex of our children except trust to nature") is an un-
fortunate but human failing. *Most* doctors are open-minded
enough to at least permit their patients to try the sex-
selection techniques. Once they examine the prescribed
procedures, they quickly realize that no harm can come of
them. Other doctors know from experience (some of it in
their own households) that considerable good can result.
Unfortunately, the ignorance of those physicians who actively discourage their patients from trying the methods
spells anything but "bliss" for their patients, many of
whom are fed up with the tired old refrain that "doctor
knows best" and that they have no more than a 50–50
chance of having a child of the sex they desire.

Consider this statement made a few years ago by Dr.
Raymond Vande Wiele of Columbia-Presbyterian Medical

Center: "Not a single convincing article has been published in scientific journals that shows you can increase the
possibility of having one sex or the other. However, it's
just a matter of years until we will be able to separate X
and Y effectively enough to artificially inseminate for a
boy or girl baby, using the husband's sperm. As for the
rest, it's all folklore. Any doctor has a 50–50 chance of being right."

Apart from considering the many scientific articles published in leading medical and science journals by Dr. Shettles and other respected researchers, some of which will be
discussed later in this book, the reader is invited to com-
pare the credentials of Dr. Shettles with those of Dr. Vande
Wiele or any other critic in such readily available sources
as *Who's Who in America, American Men of Medicine,
American Men of Science,* and the *Directory of Medical
Specialists* and then decide who is "convincing" and who
is not. (For those with a deeper interest in the conflict between Dr. Wiele and Dr. Shettles, we recommend the article entitled "The Embryo Sweepstakes" in the September 15, 1974 issue of *The New York Times Magazine.*)
As for the methods of separating the two types of sperm
in the lab, Dr. Shettles is at the forefront of current research. Some very promising developments in this field are
discussed in Chapter Nine.

But the cheapest of the "cheap shots" taken at our first
book appeared in Ann Landers' advice column, of all
places:

Dear Ann Landers,

After reading those letters from women who produced only girls and the responses from women with all boys, I'd like to know where those dames have been these last five years. Science has discovered a technique that works about 90 percent of the time. A couple can get a boy or a girl if they follow a few simple instructions. I know you can't get too clinical in your column, but it has to do with an acid condition of the female (which makes girls) and the alkaline condition (which makes boys). Why don't you tell them about it? Or maybe you are ignorant, too?

In the Know

Ms. Landers answered thus:

Dear In:

We are ALL ignorant—only on different subjects. On this subject, however, it's you. Yes, I have heard and read about the acid-alkaline theory, and it is a crock of cranberries. There is no scientific evidence to support it, according to my distinguished consultant, Dr. Roy Greep, who heads the Laboratory of Human Reproduction and Reproductive Biology at Harvard.

If Ms. Landers and Dr. Greep had really done their homework, we do not believe they could conscientiously have made those statements. In the first place, acidity and alkalinity are relatively incidental and by no means define the method we have proposed. If Ms. Landers had really read the sex-selection literature in detail and were objective in her treatment of this subject, she would first have corrected her reader on this oversimplification. We have openly presented our evidence in support of this theory and herewith are presenting considerably more. If Dr. Greep has evidence to refute the findings of Dr. Shettles and other researchers, we urge him to present it—even in Ann Landers' column. Meanwhile, those with a more technical bent can find Dr. Shettles' research findings published in such scientific medical journals as *Nature, Gynaecologia, The American Journal of Obstetrics and Gynecology, The International Journal of Fertility,* and *Fertility and Sterility* and reported upon in such respected medical journals as *Medical World News* and in lay publications such as *Time, Newsweek,* and *The New York Times.*

All too often, the innovative researcher who first attains an elusive but much sought-after goal has to suffer the detractions of the also-rans. But there is often vindication, too. Dr. Shettles' findings have found new support and new confirmation. Yet there remain people who, *without even examining the evidence at hand or without adequately addressing themselves to it,* seek to discredit that which is new, unfamiliar, "alien" and thus threatening to them or their reputations.

Even scientists who *are* sincere in their efforts to investigate another researcher's findings, in the time-honored tradition of seeking constant and reproducible results, may through carelessness or error conclude that the first researcher was wrong. Scientific discovery is not so cut-and-dried as people think. Once something has been discovered, it has to be *rediscovered,* reproduced, perhaps several times over before it is accepted. Along the way, factors such as jealousy, incompetence, and arrogance, which have nothing to do with science, may crop up and stand in the way of acceptance.

Dr. Shettles, in his more than thirty years of research, however, has lived long enough to see some of his once-doubted discoveries vindicated. In 1971, for example, he published a paper in *Nature* (March 5, Vol. 230, pp. 52–53) entitled "Use of the Y Chromosome in Prenatal Sex Determination." This paper described an ingeniously simple means of determining the sex of a fetus in the very earliest stages of pregnancy without risk to mother or child. Such a finding is of major importance to genetic counselors working with parents who can safely give birth to children of one sex but not to the other, owing to serious sex-linked genetic defects the vulnerable sex would likely be born with.

Many scientists doubted these findings. Many dismissed them out of hand, without even checking the data; others said they had tried unsuccessfully to reproduce Dr. Shettles' results in this area. Then, nearly five years later, the cover story of the December 1, 1975 issue of *Medical World News* ("Boy or Girl: Now Choice, Not Chance") stated: "Other investigators were unable to reproduce Dr. Shettles' sex-selection scores until a year ago, when a graduate student in fetology, Samuel A. Rhine of Indiana University's Department of Medical Genetics, told the Central Association of Obstetricians and Gynecologists that by using Y-body fluorescence [as had been used by Dr. Shettles] he had correctly predicted the gender of thirty-one of thirty-six babies delivered." A full report of the confirming Indiana effort, coauthored by Rhine and Drs. Jeffrey L. Cain, Robert E. Cleary, Catherine G. Palmer, and Joseph F. Thompson, has been published in *American Journal of Obstetrics and Gynecology* (May 1975).

Those who read the first book, *Your Baby's Sex: Now You Can Choose,* may be curious about its fate. Apart from going through several printings in both hard cover and paperback in this country, it has been translated into French, German, Italian, Finnish, Spanish, Japanese, Turkish, Dutch, Portuguese, Greek, Indonesian, Arabic, and some of the languages of India, where the desire for male children and the need for population control is so acute. The book has been used by the Population Council of Taiwan in their birth-control program, serialized in leading publications throughout the world, scrutinized and debated in many national and international telecasts, discussed by commentators as diverse as Russell Baker and Norman Mailer, and even recommended for use by genetics students at the University of Queensland, Australia.

Much has happened in the sex-selection field since publication of that book. Other researchers have been encouraged to enter the field, and slowly but surely confirmation of Dr. Shettles' methodology is accumulating. As expected in a field so new and so complex, there has also been some "contrary" evidence. In the pages that follow, we do not shrink from examining and rebutting the contrary evidence, nor are we modest about the confirmatory findings. Where we believe we can improve on the old, we won't hesitate to admit that what we once thought was "good enough" will no longer do.

But beyond the growing research interest in sex selection, the subject has also become a hot topic among "bioethecists," a new breed of social/biological scientists who are rightly concerned about the uses to which man is now or may soon be putting his increasingly awesome control over life forces, which were once the exclusive province of what some may call God and others may call Nature. We wish to explore not only *how to* select the sex of a baby but *whether* one should select at all. What are the possible consequences of our new ability to choose sex? Might it result in a world top-heavy with boys or brimming over with girls? If the male sex predominates, could this lead to increased crime, homosexuality, neurosis, war? Could it, as one sociologist has suggested, result in the downfall of the two-party system in the United States? Are these fears, discussed further in the last chapter, largely preposterous? What about the possible benefits?

Though we intend to look at all sides of the new bioethical debate, it should come as no surprise that we believe that the ability to choose the sex of one's offspring will provide benefits far outweighing any possible disadvantages. This time, to make our point, we call upon Abigail Van Buren, Ann Landers' generally more knowlegeable twin sister. Not long ago a woman wrote to "Dear Abby," complaining that all her children were daughters.

To begin with, my husband and I decided on only four children, but when they were all girls, he wanted a boy so much we had a fifth. When it was another girl I agreed to try just once more. Well, we got another girl. And still another and another, and now we have EIGHT girls, so I told the doctor to fix me up so I wouldn't have any more. Now I feel guilty for asking to be fixed up in case my husband asks me to try for a boy. I guess I'm writing to you because I want you to tell me that I have done my duty and shouldn't feel guilty. Will you, please?—Mother of Eight Dolls.

Abby answered:

Dear Mother: Absolutely. Don't feel guilty. And it's not too early to start saving your money for all those weddings you may have to foot the bill for.

We can only conclude that this woman—and society— would have been better served if her doctor had told her long ago that there now exists at least the possibility of *choosing* the sex of a baby. Daughters are nice—but eight of 'em? Good grief!

We hope that this book will serve both those who read our first book and those who are just now beginning to plan families and were not aware of our earlier publications. For our former readers, this book will serve as an essential update, outlining refinements in technique, presenting the new evidence in support, contending with differing theories, marshaling the latest results. For new readers, the book will provide in complete detail all you need to know in the effort to select the sex of your off-spring. Finally, for old and new readers alike, the book will examine the bioethical issues and report on a "second generation" of sex-selection techniques that are now being developed, one of which, though controversial and not widely available, already offers a virtual 100 percent guarantee of giving birth to the sex desired.

CHAPTER 2

Early Efforts Reexamined: Will Facing Away from the Sun Produce Sons?

Despite the still-common lack of concern shown by many baby doctors, few topics interest prospective parents more than the sex of their children. Records show that since earliest times, men and women have tried to choose the sex of their children, indulging in macabre rituals and superstition, even selling their souls to what we can presume were imaginary devils in efforts to beget a son or daughter. Great kingdoms have been disrupted for the simple reason that a queen—or a whole series of them— could not produce a male heir. On a less exalted level, thousands of blameless wives have been cast out of their homes because they have similarly failed to give birth to children of the sex desired by their husbands.

Today, the anguish that can result from such "failures" is as great as it was in the old days—perhaps greater, since we have come to expect so much of modern science and medicine. Parents are understandably exasperated by doctors who can do no better than offer them membership in the "50–50 Club" or tell them that Mother Nature *does* produce very nearly an equal number of boys and girls (the actual ratio in this country is 105 boys for every 100 girls). What these doctors gloss over is the fact that while this equality exists in society as a whole, it does not, unfortunately, prevail within individual family units. The fact that the four Jones boys balance the four Smith girls may content Mother Nature, but it doesn't do much to cheer either the Smiths or the Joneses.

Parents have only to look at the front page of their daily newspaper to see that this is an age of great scientific advances. Atomic energy is being harnessed for peaceful purposes, men have left their footprints on the moon, hearts and other organs are being transplanted, molecular biologists continue to unlock the deepest secrets of "inner space," even assembling genes, the units of heredity, in test tubes. Why then, many ask, is such a seemingly simple —and important—matter as the sex of one's children beyond the reach of science?

Fortunately, that is no longer the case, which is the whole point of this book. But why has it taken scientists so long to arrive at some of the conclusions and discoveries that are discussed in the following pages? The answer, in part, has to do with the traditional notion that the control of life, including the sex of our children, is the *sole* responsibility of the Creator. This notion has often served as a camouflage for ignorance, behind which the medical profession has occasionally taken refuge. In times past, when a doctor couldn't cure a disease he could justify any

resulting deaths as "God's will." In addition, for long
periods in history, anatomists and other researchers were,
for religious reasons, denied the use of the human body
for study. If they wanted to learn about the construction
and workings of the human body, they could do so only
indirectly through the study of animals, which often had
little in common with man. In the Middle Ages, for example, no one was allowed to dissect a human corpse.

It's not surprising then that over the centuries men and
women, intent upon having a say in the sex of their offspring, have been forced to resort to methods that at worst
had about them the odor of hobgoblin and humbug and
at best the faint patina of crude science. Although it has
been the men, ever eager to "prove" their virility and per-
petuate their names by begetting male children, who have
been the originators of most of these schemes, the women
have generally had to be the guinea pigs. Consider, for
example, the ordeal of some mothers in the Middle Ages:
if they wanted to please their husbands and produce boys,
they were advised by local "wise men" to drink a gamy
concoction of wine and lion's blood mixed by an alchemist
in the proper proportions. Then, while an abbot prayed,
they were directed to copulate under a full moon. When
the child turned out to be a girl in spite of these heroics,
the wise men were usually ready with an explanation; perhaps a tiny cloud had suddenly obscured the masculinizing
glow of the moon just at the moment of truth. In any
event, despite countless failures, women faithfully went on
guzzling lion's blood, and abbots fervently kept on pray-

ing. It is estimated that more than five hundred such "formulas" have been devised and recorded.

Some of our misconceptions about sex selection through the centuries can be attributed to the fact that many of the animals that anatomists studied in place of humans had two uteruses. Naturally they assumed that people did, too. Thus it didn't seem farfetched at all when Parmenides of Elea, a Greek philosopher of the fifth century B.C., announced to the world that males develop in the right chamber and females in the left. This theory soon insinuated its way into the marital bed; wives were obliged to lie on their left side if a girl was desired and on their right in the more likely event that a boy was in demand. It was believed that this would cause the husband's semen to flow toward the appropriate uterine chamber.

Things became more complicated when Anaxagoras of the same era hypothesized that it was the testicles of the male that determined sex. The products of the right testicle produced boys, while those of the left resulted in girls, according to this theory. (Note that in all these theories, it is *right* side—whether it be the right ovary, the right testicle, the right side of the body, or even the right side of the bed—that results in male offspring. Since right-handedness has always been associated with strength and justice, while left-handedness has traditionally been maligned as weak and evil, it doesn't take much imagination to understand the rationale behind these theories, all of which were concocted by men.)

Some of those who subscribed to both of these right-left

theories found themselves in some rather strange positions. They insisted that in order to beget male offspring, *both* partners had to lie on their right sides during intercourse, so that right testicle and right uterus or ovary would be perfectly aligned. A variation on this theme was the idea that a man's testicles are of different sizes, and that the larger one produces the boys. Provided the husband could figure out which was the larger, he would then take care to lie on that side. One school of sex selection at that time made life considerably harder—this time for the men—by insisting that in addition to proper positioning, the right testicle should be firmly tied with a string if a boy was wanted. It was Hippocrates, the Father of Medicine himself, who originated this knotty formula for male offspring.

Democritus and Aristotle, more than 2,000 years ago, had their own school of sex selection. According to their theories, women as well as men produce semen. If the female semen dominates, they said, the offspring will be female; if the male fluid prevails, then a boy will result. Democritus, who was a little more democratic in his outlook than some of his peers, said that while the dominant semen decided sex, a child's other characteristics were a "mosaic" formed through the intermingling of the male and female semen. A male child, he said, even though it inherited its genitals from its father, might still inherit its hands or face from its mother. In a crude way, this theory foreshadowed the modern chromosomal theory of development.

Aristotle believed that semen dominance was a direct

correlate of "vigor." The partner who is the more vigorous, particularly during the sex act, he said, determines the sex of the child. And he added that vigor was very often determined by the weather, some conditions favoring the wife and some the husband. He wrote that "more males are born if copulation takes place when a north wind than when a south wind blows, for the south wind is moister. Shepherds say that it makes a difference not only if copulation takes place during a north or a south wind, but even if the animals while copulating look toward the south or the north. So small a thing will sometimes turn the scale." One can only conclude that life was not easy for the ancients, who had to throw themselves not only right and left but north and south, too.

Later the theory of "encasement" was popular. This quaint doctrine maintained that encased within the male and female sex glands are perfectly formed individuals of microscopic size. In turn, each of these "homonuclei," as they were called, was said to contain even smaller homonuclei, and so on in an unending series representing unending generations to come. When a homonucleus became large enough, it was supposedly extruded into the womb where, though already perfectly formed, it grew to a size sufficient to maintain life in the outside world. Various formulas for the selection of sex grew up around this theory, based on complicated notions about the ways in which the homonuclei are arranged one within another.

In more recent times, pseudoscientific theories have evolved, few enjoying more of a vogue than that of E.

Rumley Dawson, a fellow of the Royal Society of Medicine, who before World War I popularized the notion that women—and women alone—are responsible for the sex of children. If the egg comes from the right ovary, he maintained, then the child will be a boy; if it comes from the left, it will be a girl. Dr. Dawson wrote countless articles and a book, *The Causation of Sex in Man,* in support of his theory, which, he said, could be used to select sex. According to his findings, the ovaries ovulate alternately, one this month and the other the next month. So after the birth of the first child, he argued, it would be a simple matter to keep track of the cycles and to time intercourse accordingly. Hence there was talk among women of "little-boy months" and "little-girl months." Apart from the fact that it has been proved that the male determines sex, Dr. Dawson erred as well in his contention that the ovaries always alternate in ovulation.

So far, many of the theories mentioned have a patina of science, but others rely on superstition and blind faith. Some reincarnation cults maintain that we simply alternate sexes in our various existences here on earth. If one is male this time around, he will be female next time—unless his soul is really set on one sex or the other; even then, according to these cults, individual choice will be denied if there is a shortage of one sex or the other.

Everywhere, it seems, boys are the preferred sex. Among the Ossets of the Central Caucasus, mothers-to-be spent their pregnancies in their home villages, separated from their husbands. If they gave birth to girls, they left the babies behind and returned to their men empty-handed, and nothing was said. But if they produced sons, they returned with their babies and heaps of gifts for the "successful" husbands. Similarly effective means of "selecting" sex have been used by a great many peoples including, at various times, the Eskimos and the Maori of New Zealand. During the early days of the British occupation, the Radshucmors of India killed nearly 10,000 newborn girls every year.

Less drastic measures were used by the young brides of the Palau Islands, who dressed in men's clothing before intercourse, believing that this would bring them male offspring. In Sweden brides-to-be slept with small boys on the eve of their weddings, again in order to ensure the birth of boys. In the Spessart Mountains of Germany, some husbands desiring male offspring still take an ax to bed with them and, during intercourse, chant to their wives, "Ruck, ruck, roy, you shall have a boy!" If they already have enough boys, the husbands dispense with the ax and chant instead: "Ruck, ruck, rade, you shall have a maid!"

In parts of Austria, some peasants believe that a year with a good nut harvest will also yield an abundance of boys. To help things along, midwives frequently bury the afterbirth under a nut tree—thus supposedly making sure that the next child will be a boy. In parts of Czechoslovakia, the bride lets a small boy step on her hands, while in southern Yugoslavia the couple takes a boy to bed with them on their wedding night, again in the effort to beget a son.

In many of the Slavic countries, the wife is directed to pinch her husband's right testicle during intercourse. In the Italian province of Modena, the husband bites his wife's right ear. And in the backwoods of Pennsylvania some men still hang their pants on the right side of the bed if they want a boy, on the left side if they want a girl.

Other folk theories maintain that boys will be conceived in the full of the moon, that the sex of the child will usually be that of the older parent, that the tides of the ocean determine sex, and that sweet foods will result in girls, while bitter or sour foods will produce boys. Some still believe that sex determination is a battle fought well into pregnancy, with male factors warring against female factors. In order to intervene in the "warfare" and tip the balance in favor of the desired sex, the expectant mother is advised to resort to a regimen of dietary "artillery," according to one folk theory. The most common prescription here calls for endless cakes and candies if one is hoping for female offspring and for "lots of good red meat" if one wants sons. It might be argued that this is at least an improvement over lion's blood.

It would no doubt be scientifically arrogant to take the position that *all* of these early attempts at sex selection were entirely foolish. Diet, for example, could conceivably have an effect; for example, note the marked effects coffee can have on sperm performance, as discussed in Chapter Six. A few years ago, if we had read that some of the ancients believed that one's clothing could possibly affect sex ratios, we would have laughed. As will be discussed later, we know now that tight-fitting clothing around the male genital area can indeed have some interesting effects on fertility and sex ratios.

Finally, there is the recent intriguing report of noted British author Roald Dahl, published in *The New York Times*. Dahl tells what he claims is the true story of one Farmer Rummins, who for years had an enviable herd of dairy cattle. In the years just after the war, Dahl lived with his mother and younger sister in a country house not far from Rummins' farm. The Dahls kept a cow in their orchard to help offset the postwar milk shortage. When the cow was ready for breeding, Dahl went to neighbor Rummins to borrow his prize bull.

Dahl was amused when Rummins asked if he wanted a heifer or a bull but, observing that it was milk not beef they wanted, said that if he had his choice he'd choose a heifer. Rummins surprised the young man by saying, all right then, a heifer he'd get. And before loosing his eager bull, the old farmer faced the Dahl cow into the sun, which was supposed to guarantee the birth of a heifer.

When Dahl expressed doubt, Rummins fumed. "Don't be so damn silly," he said. "Facts is facts."

Whereupon he led the doubting Thomas into the farmhouse and produced a stack of ledgers recording every mating carried out in the last thirty-two years. There were columns in the ledgers for date of mating, date of birth, sex of calf. Under the latter was the unvarying refrain: "Heifer, heifer, heifer. . . ." Since this was and always had been a dairy farm, Rummins said, that was what they had wanted—heifers.

When Dahl finally spotted an entry reading "bull calf"

he found next to it the notation, "Cow jumped around."
Some cows, Rummins explained, wouldn't stay faced into
the sun during the typical, very brief (often about thirty-
second) encounter with the bull—and cows that *faced
away from the sun* during intercourse, he said, always gave
birth to *bulls*. Dahl expressed amazement and asked per-
mission to go through the entire stack of ledgers. Over that
thirty-two-year period the records showed that 2,516 heifer
calves had been born, against only 56 bull calves.

A now thoroughly impressed Dahl asked the farmer why
he had never told anyone about his technique. "I reckon
it ain't nobody else's business," Rummins answered. The
procedure had made him prosperous—that was enough.

His father had passed the knowledge on to him. "He explained to me," Rummins said, "that a cow don't have nothing to do with deciding the sex of the calf. All a cow's got is
an egg. It's the bull decides what the sex is going to be.
The sperm of the bull. According to my old dad, a bull
has two different kinds of sperm, female sperm and male
sperm. . . . So when the old bull shoots off his sperm into
the cow, a sort of swimming race takes place between the
male and the female sperm to see which one can reach the
egg first. If the female sperm wins, you get a heifer."

The old farmer's father was correct in all of this and
surprisingly far ahead of the science of his time, another
reason why we should not be too quick to dismiss his ideas
as nonsense. The farmer's theory was that the sun exerts a
pull (like that of the moon on ocean tides), which for some
reason makes the female-producing sperm swim faster.

"And if you turn the cow around the other way," he

said, "it's pulling them backwards and the male sperm wins instead."

Dahl wondered if the technique might work on humans, as well. Rummins said it certainly would, "Just so long as you remember everything's got to be pointed in the right direction. A cow ain't lying down, you know. It's standing on all fours. And it ain't no good doing it at night either because at night the sun is shielded behind the earth and it can't influence anything."

When Dahl asked if the farmer had any proof that the procedure might work as effectively with people, the old man responded with "another of his long sly broken-toothed grins" and the challenging statement: "I've got four boys of my own, ain't I? Ruddy girls ain't no use to me around here. Boys is what you want on a farm."

Whether there is more bull in this story than meets the eye we don't know. If those records really exist as repre-sented, then some very interesting phenomenon was at work on the Rummins farm, one that is worthy of scien-tific investigation. Just because something seems odd or unlikely is no reason to dismiss it out of hand. Our prede-cessors were not always as "stupid" or "superstitious" as we sometimes like to think. At any rate, we do not wish to be accused of the smug scientific "know-nothing" at-titude that some of our colleagues so casually take today. We're recommending neither lion's blood nor facing into (or away from) the sun in broad daylight—but if by chance you try it and it works, we'd be hurt if you didn't tell us about it!

CHAPTER **3**

Sex Selection the Natural Way:
Understanding the Basics

Man's battle against Mother Nature has not been an easy one. Nor was it very successful until man began to turn his attention away from external forces (north winds, full moons, tides, and nut trees) and began focusing, instead, on *internal* factors within his own body. True enough, the ancient Greeks and a few others had come up with theories based on biological considerations, but these were predicated more on guesswork, prejudice, and wishful thinking than on scientific observation.

Scientists began to make real strides forward in attempts at sex selection in 1827 when Karl von Baer identified the mammalian ovum or egg that is produced each month by the female ovary. Working with dogs, this pioneering biologist traced the development of the tiny embryos in the wombs of his laboratory animals back to the moment of conception. In the course of this, he discovered that before the embryo—the combination of sperm and egg—attaches itself to the lining of the uterus it passes through one of the two Fallopian tubes that open into the uterus. He was not certain that the embryo had not simply been ejaculated into the female by her mate during intercourse until he examined one of the ovaries at the end of the Fallopian tube and discovered a bulging follicle on its surface. Opening this, he discovered a tiny yellowish speck, which under the microscope turned out to be identical in appearance to the "ovules" he had previously observed in the Fallopian tubes.

It wasn't until 1841 that the anatomist Rudolf von Kölliker conclusively established that sperm originated within the tissue of the male testes. This discovery was quickly followed by observation in animals of the actual penetration of the egg cell by the much tinier sperm cell. The real nature of fertilization in these studies, however, didn't become apparent until the 1870s, when a series of researchers concluded that the sperm and the egg combined in such a way that each contributed about equally to the resulting embryo. In 1883, Pierre van Beneden, another European scientist, showed that this was precisely the case, when he discovered that the central portion, or nucleus, of both egg and sperm contain only half the number of chromosomes that are present in the nuclei of all our body cells.

Chromosomes are the microscopic, rodlike structures that contain the even tinier genes, which determine the color of our eyes, our facial features, the color of our skin, and all other bodily characteristics. Every individual is made up of body cells and sex cells. The sex cells are the sperm or the eggs (ova). All the other cells—those that make up bone, skin, flesh, and so on—are called body cells. In man, each body cell contains forty-six of these rodlike chromosomes. Van Beneden's important discovery tells us that half of those chromosomes come from our mothers and half from our fathers, when their sex cells combine.

This discovery, however, still did not tell us just how Nature determines sex. We didn't know, for example, whether it was one of the mother's chromosomes or one of the father's chromosomes or a combination of the two that determined sex. The first suggestion that there might be *two types of sperm*—one carrying a female-producing chromosome and the other a male-producing chromosome —came in 1890. Microscopic studies of eggs indicated that the pairs of chromosomes within their nuclei were all perfectly matched. But similar studies showed that the sperm possessed *one* imperfectly matched pair.

One of the chromosomes in this odd pair was smaller than the other. Researchers quickly seized on this discrepancy as a possible explanation of the sex-determining mechanism. An American zoologist, C. E. McClung, in 1902 was the first to suggest that this seemingly mismatched pair constituted the sex chromosomes. It soon became fashionable to refer to the smaller of the two as the Y chromosome and to the larger as the X chromosome. When the sperm cells underwent maturation and divided in half, it was hypothesized that the Y went into one of the newly formed sperm cells and the X into the other.

It should be understood that all of this work was with animal cells which, apart from the fact that they were readily available with no taboos attached, are also larger than human cells and therefore easier to study. Nonetheless, the principles these researchers developed were entirely sound and turned out to be as applicable to man as to animals. It was a study of the lowly mealworm that first demonstrated that *the smaller Y chromosome produces males.* This came to light in 1905 when Dr. N. M. Stevens observed that half of the mature mealworm sperm cells contain ten large chromosomes, while the other half contain nine large chromosomes and one small one (the Y chromosome). The ova all contain ten large chromosomes. Ova fertilized by sperm containing the Y chromosome all developed into males, while those fertilized by sperm containing the X chromosome yielded females without exception.

Momentous as this confirmation of McClung's theory was, decades passed before it became clear that the sex of humans is similarly determined. Human chromosomes are so small that it was not until 1956 that their number was firmly established at forty-six for each body cell and twenty-three for each sex cell. Indeed, no one had ever actually seen a human egg cell undergoing fertilization by a sperm cell until the latter part of the 1940s and the early 1950s, when Dr. John Rock of Harvard and Dr. Landrum B. Shettles of Columbia watched the miracle of conception on the stages of their microscopes. And although they could see the sperm penetrating the egg, they could not, in these living specimens, visualize the chromosomes themselves. They knew, however, from studies with stained and fixed specimens, that the X and Y chromosomes were there and that they determine sex. We shall see in the next chapter that Dr. Shettles has since discovered a means of visually identifying the X-carrying and the Y-carrying sperm cells in living specimens.

Let us pull together all the information that science had gathered about human fertilization and sex determination by 1960 and put it to use in describing the conception of a baby. Will it be a boy or a girl? Follow along and we'll see:

First, the prelude to our drama, the preparation of the sperm and the eggs. At birth, a baby girl's ovaries contain more than half a million egg cells. That, of course, is far more than she will ever need. In fact, not more than five hundred of these will mature during her lifetime to be released—one each month—from her ovaries. Why there are so many remains one of the mysteries of medical science.

Generally, though not always, the process of ovulation or egg production alternates between left and right ovary on a monthly cycle. Usually a week to twelve days after a woman's menstrual period, a tiny follicle (which simply means "little bag" in Latin) erupts from the surface of the ovary. This watery blister bursts open—sometimes causing the woman to feel a sharp twinge of pain—and the egg that it contains tumbles out and into the clutches of the lacy fimbriae or "fingers" of the Fallopian tube. These finger-like structures draw the egg, which is no bigger than the point of a pin, into the tube. At this point the egg, the nucleus of which resembles the yoke of a chicken's egg in color and shape, is encased in a gelatinous mass of 4,000 to 5,000 "nurse" cells, which nourish and protect it during its first vulnerable hours outside the ovary.

As the egg bounces gently down the four-inch-long Fallopian tube, propelled by hairlike cilia that wave like tall grass in a soft breeze, the nurse cells gradually slough off and are dissolved by enzymes. At this point the placid egg is ready to meet her mate, the sperm cell, which is anything but placid.

While the egg is more than $\frac{1}{250}$ of an inch in diameter, the main body of the sperm is a mere $\frac{1}{8000}$ of an inch across, and its volume is only $\frac{1}{50,000}$ that of the egg. They are the smallest cells in the body. It has been estimated that all of the sperm necessary to produce the next generation in the United States could be contained in the space of a pin head, while the eggs necessary for the same job would fill a pint jar. During intercourse, the male ejaculates about 400 million sperm cells into the vagina. Why does the male produce and release so many of these microscopic creatures? Here we do know the answer—or at least part of it. It is because the vaginal environment is so hostile to the sperm cells, which die off by the millions shortly after they are released, slaughtered by the acid that abounds in the vagina.

Sperm cells resemble tadpoles with their rounded heads and long tails, which they use for propulsion, speeding (and considering their size, that's an accurate description) through the vaginal and cervical secretions at a tenth of an inch per minute. Again taking their size into account, the seven-inch journey through the birth canal and womb to the waiting egg is equivalent to a five-hundred-mile upstream swim for a salmon! Yet they often make this hazardous journey in under an hour, more than earning their title as "the most powerful and rapid living creatures on earth."

Only the fittest survive to pass through the cervix into the womb. Here they find a more hospitable environment, more alkaline than acidic. Still, many die along the way; others smash into the back of the womb or go up the wrong Fallopian tube. Many of those that go up the right one will miss the egg anyway, if only by a millionth of an inch. The idea that the egg exerts some magical power of attraction was disproved under Dr. Shettles' microscope. Those that reach the egg—and thousands of them do make it—hit it blindly. Soon the egg looks rather like a pincushion, except that in this case the "pins" beat their tails furiously, trying to drill into the egg. That is a sight never to be forgotten, one that Dr. Shettles calls the "dance of love" (see Figure 1).

Under the microscope one can see the sperm making heroic efforts to gain admittance to the egg's inner sanctum, which houses the nucleus and the chromosomes. Many are able to break through the egg's outer core, *but only one penetrates the interior,* tail and all, there to merge with the egg's nucleus and create a new human being. As soon as one sperm penetrates the nucleus, all others find the way blocked. Some unexplained mechanism within the egg apparently releases a chemical that renders the innermost portion absolutely impregnable once it has been fertilized by a single sperm. The egg's unsuccessful "suitors" wear themselves out "pounding at the door" and finally die of exhaustion.

As you will recall, the sperm carries twenty-three chromosomes and so does the egg. Twenty-two of these (in each) match up as pairs that determine all the bodily characteristics of the new individual—except for sex. The two remaining chromosomes decide the subject's sex. The female *always* contributes an X chromosome. If the sperm that penetrates the ovum also carries an X chromosome, the resulting individual will be XX, otherwise known as a girl. But if the sperm carries the Y chromosome, the baby will be XY which, to the geneticist, spells b-o-y.

And that's how Mother Nature does it.

CHAPTER 4

Boy or Girl? Critical Discoveries upon Which the Method Is Based

On June 5, 1960, the following headline appeared over a full column of type in *The New York Times:*

2 SHAPES OF SPERM
FOUND IN HUMANS

Discovery of Distinct Types
Suggests Clues to Study
of Sex Determination

The story attached to the headline is reprinted here in full:

Human spermatozoa take two distinctly different shapes, a specialist at Columbia University's College of Physicians and Surgeons has reported.

The difference may show which sperm will produce male offspring and which will produce females, the evidence suggests. If this is supported by further research it will provide the first means ever found of identifying the two types visually. It might make it possible to predict whether a prospective father is more likely to produce a boy or a girl.

It might also lead to an effective way of controlling the outcome.

It is generally accepted that the spermatozoon, the male germ cell, rather than the female egg, determines whether the offspring will be a boy or a girl.

However, structural differences indicating which individual sperm will produce each effect have not previously been known. Textbooks describe normal human spermatozoa as tadpole-shaped and fairly uniform in outline.

Dr. Landrum B. Shettles, assistant clinical professor of obstetrics and gynecology at Columbia University, however, finds two distinct populations.

In one the sperm head is round and compact. In the other it is larger, elongated and oval—somewhat football-shaped. There appear to be no intermediate types.

An initial report on Dr. Shettles' research was published in the May 21 issue of *Nature,* the British science magazine. Forty dried and unstained sperm specimens were studied under the phase-contrast microscope. Continuing research has expanded the number studied to more than 100 from different persons.

Two items of evidence suggest that the sperm with elongated heads are the potential producers of girl babies while the round-headed ones carry the genetic instructions for the production of a boy.

One clue was found in the observation that in sperm specimens from most of the donors studied so far the round-heads were substantially more numerous than the oval types.

This fact may possibly explain why more male than female babies are conceived and born in a normal human population.

The other clue concerns the interior structure that Dr. Shettles believes can be discerned in the specimens.

He and other observers have been able to distinguish nuclear structures in the sperm heads that appear to be recognizable as chromosomes. In one sperm head three observers were able to count eighteen distinct chromosomes, only five fewer than the total number the sperm is known to contain.

The chromosomes are the carriers of genetic information from one generation to the next. Each parent contributes a complete set to the child. The interaction of those sets determines all manner of characteristics including size, eye color and sex.

In the determining of sex, the female can provide only the chromosome compatible with development of another female. Geneticists call this the X chromosome. The male can contribute either an X, in which case the offspring, having two X chromosomes, will be female, or a Y, which will pair with the female's X to produce a male.

Dr. Shettles' studies showed that the most centrally located chromosomes discernible in the sperm heads possessed the exact size and shape relationships known to be associated with the X and Y chromosomes.

The sizes and shapes indicate that the round-headed sperm carry the Y chromosome while the somewhat larger oval sperm heads contain the X, the report said.

Attempts here and abroad have been made to separate male from female-producing sperm in a given specimen by electrical and other means. The efficiency of these attempts has been debatable, Dr. Shettles said.

The ability to tell the two apart by microscopic observation would be a valuable method for checking the effectiveness of such separation, the article in *Nature* said. Further research might provide a still more effective way of making such a separation and lead to the possibility of controlling the sex of the offspring.

The research is continuing. Two further and more detailed articles on Dr. Shettles' discoveries have been accepted for publication in scientific journals in this country.

It is hardly surprising that it was Dr. Shettles who made this discovery. In the 1950s he had stunned much of the scientific world by repeatedly achieving "test-tube conception"—the fertilization of human ova by human sperm cells *outside the womb,* a technical feat that was not duplicated by other scientists for more than a decade. And though there was opposition to this work, Dr. Shettles believed it was necessary and of the utmost importance if we were ever to fully understand human embryology. Thanks to his studies and the microphotographic records he made of all that he observed, medical students were able for the first time to study directly the development of a human being from the moment of conception onward. Today Dr. Shettles' micrographs appear in his now classic "biological atlas," *Ovum Humanum;* in more than fifty textbooks on embryology, biology, and genetics; and in greatly enlarged form in the Museum of Science in Moscow.

In conversation (with Rorvik) Dr. Shettles recalled the first discovery that led to the sex-selection procedures he now advocates.

"Medical science had known for some time," he said, "that it is the male who determines the sex of the offspring.

The man who leaves his wife because she brings him nothing but girls or nothing but boys is only kidding himself. If the man's fertilizing sperm carries an X chromosome, the child will be a girl; if it carries a Y it will be a boy."

The trouble is, he noted, doctors have never been able to tell the difference between "male" sperm and "female" sperm. About all that was known was that the Y chromosome is smaller than the X. Dr. Shettles had long felt that this difference should be reflected in the overall size of the sperm heads. With ordinary microscopy, however, killed and permanently fixed sperm specimens failed to reveal the presence of two distinct sperm populations.

"Then one night," Dr. Shettles continued, "I decided to examine some *living* sperm cells under a phase-contrast microscope." The technique of phase-contrast microscopy, which was relatively new at that time, throws eerie halos of light around dark objects, revealing details that ordinary microscopes miss. On the stage of the microscope the living sperm cells flashed through the field of vision like luminescent eels from the darkest depths of the ocean (see Figure 2). Dr. Shettles put them into slow motion by introducing a little carbon-dioxide gas into the specimen. The results were almost as electrifying as the "charged" sperm cells themselves: *almost immediately Dr. Shettles noticed that the sperm came in two distinct sizes and shapes.*

"I was so excited," he recalled, "that I ran upstairs and grabbed the first lab technician I could find. I had to show somebody what I'd found."

Now, after examining more than five hundred sperm

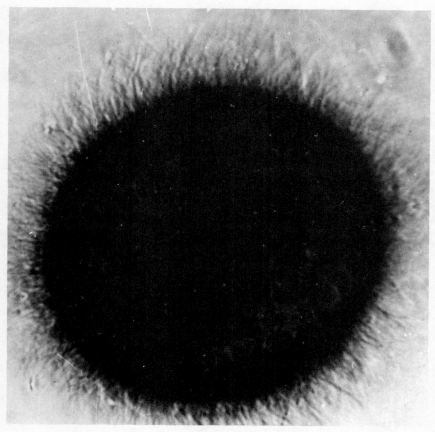

Figure 1: THE DANCE OF LOVE
Thousands of sperm, looking like pins in a pincushion, fight for admission
to the egg's inner sanctum. Only one will make it.

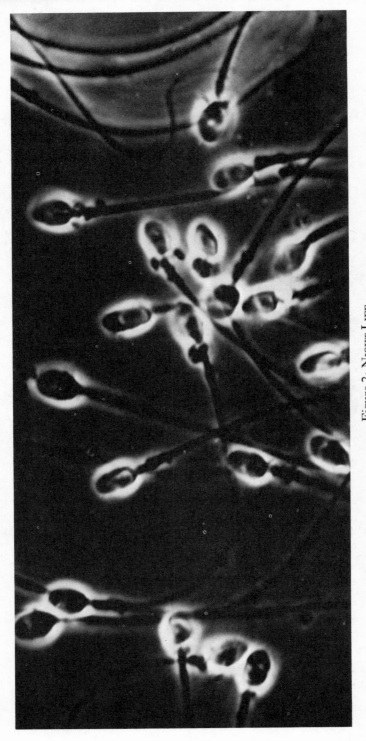

Figure 2: NIGHT LIFE
Under the phase-contrast microscope, sperm cells are shrouded with surreal halos of light.

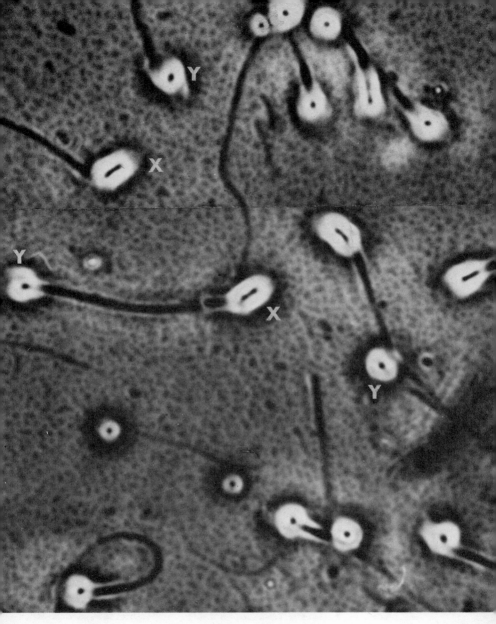

Figure 3: Boy or Girl?
X marks the female, Y the male.

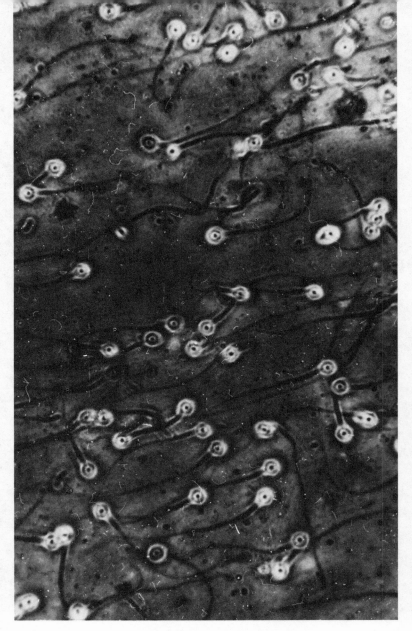

Figure 4: AND NOT A GIRL IN SIGHT
The donor of this sperm sample, populated by nothing but Y's,
comes from a family that for 256 years has produced almost nothing
but boys. Such cases are extremely rare.

specimens in the same way and with the same results, he is convinced that the two sizes correspond to the two sexes: small, roundheaded sperm carry the male-producing Y chromosomes, and the larger, oval-shaped type carry the female-producing X chromosomes (see Figure 3). He noticed that in most cases the round sperm far outnumbered the oval-shaped sperm, which is compatible with the fact that 110 to 170 boys are conceived for every 90 to 100 girls, and that for every 100 female births there are about 105 male births.

In terms of longevity, resistance to disease and stress, adaptability to environment, and so on, it has long been conceded, at least by scientists, that the male is the weaker of the two sexes. The male may have more muscular strength than the female, but he lacks her staying power. This fact now appears to be borne out even at the most elemental level; the male-producing sperm begin with a substantial head start in terms of numbers (perhaps a two to one margin, because of an extra cell division), but end up only slightly ahead of the female-producing sperm.

With this much to go on, Dr. Shettles began checking the family histories of the men whose sperm he had examined. He failed to find anyone who produced only the oval-shaped female sperm, but he did encounter some specimens which contained almost nothing but the round-headed variety. In each of the latter cases, his physiological sleuthing revealed a man who had produced nothing but male offspring. In cases in which the longheaded sperms prevailed, he generally found fathers surrounded by little girls (and wives who wanted boys). Dr. Shettles stressed here, however, the rarity of cases in which the husband produces sperm that is predominantly of one type. And even in cases in which a man may produce considerably more sperm of one type than the other *he can very often still produce offspring of both sexes,* provided he follows certain procedures that we will discuss later. (Just why, in these rare cases, some men "favor" one type of sperm over the other is still not understood, though genetic factors seem to play an important role. The point to remember is that a strong predisposing factor, genetic or otherwise, to produce one sex or the other is absent in most males, and even where present, it can often be overcome. Individuals who have repeatedly fathered children of the same sex are more often the victims of bad luck than of genetics.)

In his laboratory, Dr. Shettles fondly flicks through the pages of his scientific papers, pointing to pictures of his patients' sperm.

"What do you notice about this one?" he will ask, and you must acknowledge that it contains "nothing but roundheaded ones" (see Figure 4). And then he explains that its donor comes from a family that for 256 years has produced almost nothing but male offspring, the "almost" being reserved for two females that came along during those two and half centuries. Then Dr. Shettles asks you to examine yet another micrograph. You remark that this one is populated with both types of sperm in fairly equal numbers, at which the doctor admits that the specimen is his own and that he is the father of four girls and three boys.

After making his discovery, Dr. Shettles published his findings in the prestigious scientific journal *Nature* and suddenly found himself in a new controversy. Lord Rothschild, a British physiologist, apparently suffering from the well-known "British First" syndrome (British scientists desperately try to beat their American counterparts to significant discoveries), went on television to characterize Dr. Shettles' findings as "a lot of tripe." Lord Rothschild attempted to support his evaluation with a curious list of objections, which to some observers seemed only to reveal Lord Rothschild's lack of expertise in dealing with phase-contrast microscopy.

Perhaps the best account of this early controversy appeared in the August 12, 1960, issue of *Medical World News*. The headline announced: "Sperm Shape Study Whips Up a Storm—In a trans-Atlantic controversy, two noted genetics investigators debate the validity of microscope evidence on the role of sperm shape in determining sex of offspring." The story is reprinted below:

Asked to comment on the report by Columbia University's Dr. Landrum B. Shettles that the shape and size of a sperm's head may decide sex, Britain's testy Lord Rothschild, a leading gametologist, snapped in a recent television interview: "It's a lot of tripe. Many measurements have been made in hopes of identifying X and Y spermatozoa. No such difference has been found."

Now, in more scientific terms, Lord Rothschild has detailed his criticism. Here are highlights of his charge—and Dr. Shettles' reply (in italics)—published in the British journal, *Nature:*

Lord Rothschild states:

Shettles reports that human sperm heads fall into two distinct populations without intermediate types. One type of sperm head is small and contains a centrally located round mass [Y-chromosome], while the other is large and contains a centrally located elongated mass [X-chromosome]. According to Shettles, these centrally located masses are nuclear material, though nuclear material is generally thought to be uniformly distributed within the mature sperm head.

Dr. Shettles replies:

More than a hundred men have now been studied, and the two distinct populations have been repeatedly observed in every semen specimen—regarding head and nuclear size and shape; the position, size and shape of the most central chromosomal mass; and the pattern of reflected light.

In one sperm head, it is reported, three investigators independently counted eighteen discrete chromosomes. This is the first time that chromosomes have been "observed" in a mature sperm head, it having been universally agreed before that discrete chromosomes are not visible there.

The most central mass in each type of head is unequivocally Feulgen-positive, as well as the other compact chromatin masses considered to be autosomes. Twenty-three chromosomal masses have been counted in some of the rounded [Y-chromosome] type.

Although X and Y spermatozoa might have been expected to be present in approximately equal numbers, Shettles makes no reference to having counted the types, which is surprising considering the ease with which the two types could have been counted.

Contrary to previous teaching, the ratio of the two types of heads varies from man to man; in most specimens thus

*far studied, the rounded type predominates. This problem
is being studied regarding sex of offspring possibly already
produced, frequency of emission and age of individual.
Any difference in migratory rate of the two types is also
being investigated.*

One photomicrograph, described by Shettles as being
of a human sperm head, shows that the head in question
is no less than 30 microns long, whereas the average human sperm head is 5 microns long.

*The final enlargement of the illustration being enlarged
nearly 10,000 times, and its reduction by the printer, do
not permit one to figure the original size of the head in
microns. The length of human spermatozoon heads is
often cited to vary from 5 microns to 8 microns; in fact, in
some men studied, the larger, normal elongate types had
heads 15 microns long.*

From what has been said, Shettles' claim should clearly
not be accepted. The question at issue is: What, in reality,
did he observe? Were the round and long black masses
distorted vacuoles which occur inside human sperm heads
—distorted because of the unusual optical system Shettles
used, or because, when dry, human spermatozoa just do
not always settle on the microscope slide in the same way?
Or were the central masses optical artifacts which can
easily be demonstrated if crystals in dried human seminal
plasma are examined with phase contrast at different focal
depths?

*At least 25 different competent observers have been
shown preparations; each individual has agreed regarding
the two populations. It has also been found that the two
populations can be seen in dried smears viewed in bright
light microscopy, with proper adjustment in light intensity. Moreover, the differences in morphology have been*

noted in very thin smears of seminal fluid with living spermatozoa immobilized by means of an atmosphere of carbon dioxide. Furthermore, thin, wet smears observed before, during and after drying under the phase contrast objective of the Zeiss microscope show the two, distinct types of heads. Again, the Feulgen test is confirmatory.

Concludes Rothschild: There is so far no evidence that physical differences have been found between X and Y spermatozoa.

Concludes Shettles: *There is an exact pattern of arrangement of the Feulgen-positive, concentrically arranged chromosomal material around the most central chromosome in the smaller, rounded type of head, and an ovoid pattern in the more elongated type. Two most dependable cytogeneticists have already repeated and confirmed the findings reported.*

Lord Rothschild was not heard from on this subject again.

Meanwhile, Dr. Shettles was far more interested in the implications of his research than in any controversy it might excite. He was intrigued by the possibility that "further research," as *The New York Times* had put it, "might lead to the possibility of controlling the sex of the offspring." He had but one objective in mind: to find some means of exploiting this new knowledge to help parents choose the sex of their children. As he put it, he was tired of telling patients that they would just have to be content with what Nature decided to give them. But *how* to exploit this knowledge? Well, since there definitely seemed to be a difference in the overall size of the two types of sperm, he reasoned, there must be other differences as well. Perhaps one type was stronger than the other or faster—or both. Perhaps one type could survive longer in a certain environment than could the other. There were all sorts of intriguing possibilities that could lead to a means of selecting sex—simply by interfering, even slightly, with the environment in which the sperm seeks out the egg.

From the start it seemed fairly certain that the larger, female-producing sperm (now called gynosperm) must be more resistant than the other type. Otherwise, why should there be nearly twice as many of the smaller, boy-producing variety (known as androsperm) in the ejaculate of the average male? Why, in other words, was there this apparent extra division, and hence doubling, of the androsperm, if not to compensate for some inferiority in coping with the environment beyond the male reproductive tract? This "inferiority," as suggested earlier, is borne out all through life: there are far more stillbirths and miscarriages of male children than of female; more boys than girls die during infancy; women are more resistant than men to numerous diseases, have longer life spans, and so on. No wonder the androsperm require a substantial head start—even then the boys end up only slightly out in front of the girls in terms of the number born each year.

What accounts for the greater slaughter of androsperm within the womb? To find out, Dr. Shettles began studying the environment that exists inside the vagina and uterus at about the time of conception. He took transparent capillary tubes and filled them with cervical and vaginal secre-

tions. Then he turned millions of sperm loose at the opening of the tubes and watched their activity through his
microscope.

"It was a little like watching the races at Belmont," he
said, calling his capillary tubes "physiological racetracks."

When the secretions in the tubes were more acidic than
alkaline, the gynosperm seemed to prevail. Their greater
bulk seemed to protect them from the acid for much longer
periods than their little brothers were able to survive. But
when the tubes were filled with cervical mucus that had
been removed from a woman very close to the time of
ovulation, the smaller androsperm were clear-cut winners
every time. Why? Because the ovulatory mucus is highly
alkaline, an environment that favors the androsperm.

Now to clarify: acid inhibits both gynosperm *and* androsperm, but it harms the androsperm first and most, cutting
them out of the herd and thus out of competition. Alkaline
secretions are kind to both types of sperm and generally
enhance the chances for fertilization. (That is why a
woman's body chemistry becomes increasingly alkaline as
the moment of ovulation approaches.) But in the absence
of hostile acids the androsperm are able to use the *one* advantage they have over their sisters: the speed and agility
that their small, compact heads and long tails give them.

Here Dr. Shettles had some very important information
indeed. As a gynecologist he knew that the environment
within the vagina is generally acidic, while the environment within the cervix and uterus is generally alkaline.
And he knew that the closer a woman gets to ovulation

the more alkaline her cervical secretions become. He knew, also, as a result of his "racing" experiments, that female-producing sperm are likely to outlast their male-producing counterparts in an acidic environment and, as a result of their strength and resistance (some would simply say their tough hides), manage to fight their way to the egg first. (He checked out this resistance theory in other ways; by heating the capillary tubes, for example, he was able to observe the greater staying power of the gynosperm. All of the sperm cells died from the heat eventually, but the androsperm died first.) In an alkaline environment he knew that the androsperm were clearly superior, able to use their agility to the fullest.

All of this told him that *timing of intercourse* is a critical factor in choosing the sex of children. His findings suggested that intercourse at or very close to the time of ovulation, when the secretions are most alkaline, would very likely result in male offspring. Intercourse two or three days before the time of ovulation, on the other hand, when an acid environment still prevails, would be likely to yield female offspring.

Already some readers, concerned about pinpointing the exact time of ovulation, may be wondering about the effects of frequent intercourse, starting about three days before the suspected time of ovulation. Could this "shotgun" approach to the timing problem be a means of begetting a boy, since intercourse would occur at least once at or near the time of ovulation? The answer is almost certainly negative, for this reason: gynosperm from the first intercourse in the series could, because of their superior staying power, still be lurking in the Fallopian tube three days after being deposited there, but it is very unlikely that there could still be any androsperm alive or in condition to fertilize the egg. Subsequent intercourse would probably be futile, since fertilization—by a girl-producing sperm—would already have taken place by the time male reinforcements arrived. The timing factor and the means of pinpointing ovulation are described in more detail later.

Certain now that he was on the right track, Dr. Shettles began looking through scientific and historical literature for further confirmation of his findings. He found that Orthodox Jews produce a significantly greater percentage of male offspring than does the general population. To find out why this might be so, he began consulting rabbis and poring over the Talmud, a compilation of Jewish beliefs, laws, manners, and customs that was completed between the fourth and sixth centuries A.D. One of the passages he found was this: "The determination of sex takes place at the moment of cohabitation. When the woman emits her semen before the man [meaning when she experiences orgasm before her husband], the child will be a boy. Otherwise, it will be a girl." If a boy was desired, the Talmud directed the husband to "hold back" until his wife experienced orgasm. Before considering this further, let us go on to the second point that Dr. Shettles found in Orthodox Jewish law: that women must not engage in intercourse during their "unclean" period (menstruation) or for one week thereafter.

Both of these directives coincided very neatly with Dr. Shettles' findings, explaining why Orthodox Jews have more male offspring than the rest of the population. Orgasm is the less important of the two factors, but it can play a part in sex selection; female orgasm, Dr. Shettles has found, helps provide additional alkaline secretions. Of course, some women never experience orgasm. These women should not be alarmed nor think that their chances to conceive boys are diminished. Female orgasm is probably desirable, but it is certainly optional, for there are other ways, as we shall see, of increasing the alkalinity that favors male offspring. The other point—abstaining from intercourse until at least a week after the conclusion of menstruation—is more significant, for this puts coitus at a point very close to the time of ovulation in most women, the point at which the secretions are most alkaline. And so it is, Dr. Shettles concluded, that Orthodox Jews beget many male offspring.

Dr. Shettles also sifted through the data on artificial insemination (a technique that accounts for several thousand live births each year in this country). He knew that doctors specializing in artificial insemination try to pinpoint the time of ovulation in their patients so that fertilization can be achieved on the first try, if possible. This is simply more economical and less traumatizing for the patient. It occurred to Dr. Shettles that an unintended side effect of this practice ought to be an abundance of male offspring. In a series of several thousand births achieved by artificial insemination, he found that the sex ratio was 160 boys for every 100 females. In another series, 76 percent were boys, and 24 percent were girls!

Elated that his hunch seemed to be correct, Dr. Shettles began startling some of his patients by telling them that they no longer had to rely on the whims of Mother Nature —at least not entirely—when it came to the sex of their children. He also began working at about this time with Dr. Sophia Kleegman, a professor of gynecology at New York University's Medical School and director of its Infertility Clinic. Dr. Kleegman, now deceased, was long a leader in the field of artificial insemination. She sent samples of her donors' sperm to Dr. Shettles for analysis under the phase-contrast microscope. Dr. Kleegman, as well as Dr. Shettles, achieved a high rate of success in helping her patients choose the sex of their offspring.

With all of this information in hand, Dr. Shettles believed he had put down the foundation for a workable sex-selection technique. He subsequently summarized his findings in the *International Journal of Gynaecology and Obstetrics,* September, 1970. Excerpts follow:

The difference in shape and size, as well as the correspondence of the overall ratio of sperm type with the conception rate by sex, suggest that other factors are operating, as well as pure numbers. Speed is one such factor and would seem to favor the smaller Y-bearing sperm. Since these are of less mass than the larger X-bearing sperm, they should be able to migrate through the reproductive secretions at the time of ovulation at a greater speed with the same amount of energy, thus making one of them more likely to effect fertilization. When tested in a capillary tube filled with ovulation cervical mucus over a distance of one foot, the small-headed sperm invariably wins the race.

Continence [meaning abstention from intercourse] or lack thereof is another factor which could favor one or the other type, depending on the circumstances. . . . Continence is associated with an increased frequency of round-heads [androsperm]. Oligospermia [low sperm count] is associated with female offspring. In men with sperm counts of 20 million cc and under [20 million sperms per cubic centimeter of fluid], the likelihood of female offspring varies inversely with the count. With a sperm count of a million or less, only female offspring resulted. . . . This is indicative of the X-chromosome-bearing sperm [female-producing] being the survival of the fittest.

A third factor is longevity, which seems to favor the X-bearing sperm. When the egg is ready for fertilization, this factor may be unimportant, but it is possible for fertilization to occur by a robust sperm which has survived over a period of days within the tube.

Interrelated with the above is differential environment within the cervix before and at time of ovulation. At the time of ovulation the cervical mucus is, among other things, most abundant, most alkaline, of lowest viscosity, and most conducive to sperm penetration and survival. In contrast, the more acid environment within the cervix until a day or so before ovulation is unfavorable to sperm. During this time only the more fit (female) sperm have a chance for survival. The potential to have male and female offspring obviously varies greatly among men. Utilization of each lot of reproductive talents, so to speak, is governed greatly by the timing of coitus in relation to ovulation.

These findings became the basis for the sex-selection methods we describe in more detail in the following chapters.

CHAPTER 5

New Findings, Controversy, and Confirmation

Though the basic method described in our first book remains largely intact, continuing research has yielded refinements that we believe further enhance the chances of successfully choosing the sex of your next child. Several new recommendations and procedures are described in detail in Chapters Six, Seven, and Eight; these should be studied closely. In this chapter, meanwhile, we seek to bring the reader up to date on research around the world, which in some instances strongly supports and in others seems to detract from the Shettles method. Both categories must be dealt with.

The report of Dr. Rodrigo Guerrero of the Universidad del Valle in Cali, Colombia, in the *New England Journal of Medicine* (November 14, 1974) partially supported and was partially at variance with the methodology proposed by Dr. Shettles. Like Dr. Shettles, Dr. Guerrero found that a decisive majority of the offspring conceived via artificial insemination (where care is taken to introduce the sperm at or very near the time of ovulation) are boys. But then, in conflict with Dr. Shettles' findings, Dr. Guerrero observed that "in natural insemination, the proportion of male births diminished from 68 percent six or more days before, to 44 percent on the day of the shift [in basal body temperature, signaling ovulation]."

We have concluded that Dr. Guerrero's raw data on conceptions achieved by artificial insemination are sound, inasmuch as they were the product of medical professionals working in infertility clinics in several U.S. cities where record-keeping is meticulous. But we cannot be so confident about the raw data upon which he bases his conclusions about the *natural* inseminations. For here Dr. Guerrero had to rely, retrospectively, on charts kept by women who were principally trying to avoid pregnancy through the rhythm method of birth control. The reliability of such a procedure in definitively determining the interval between last intercourse and conception appears to us very doubtful, certainly insufficient for authoritatively describing any relation between time of menstrual cycle and sex ratio. Further doubt is cast upon the Guerrero procedure by virtue of the fact that his "experimentals" were all failures in the rhythm method of birth control; it seems likely that the menstrual pattern of this population is more highly atypical than that of the general population. Thus the likelihood that ovulation was truly pinpointed in most cases appears hypothetical at best.

The statistical sampling related to conceptions that occur "six or more days before ovulation," moreover, is necessarily so small (simply because it is almost impossible for sperm to survive that long in wait of an egg to fertilize) that we are at a loss to make much sense of an evaluation based on such ethereal and very possibly artifactual observations. Certainly, as has been noted in an editorial in the same journal that published Dr. Guerrero's findings, even if those findings proved reliable, little use could be made of them, since it would take five years or more on average before a live birth could occur when one consistently halted intercourse six or more days before ovulation. Dr. Guerrero hypothesized that the difference in the results between natural and artificial insemination might be explained by the different environments in which the sperm reside prior to their introduction into the female reproductive tract. We believe the differences are more likely attributable to weaknesses in Dr. Guerrero's methodology.

It should be understood, moreover, that Dr. Guerrero and some of those who have commented on his findings (including a study group at George Washington University), have completely overlooked the fact that the Shettles method is in no way synonymous with "natural insemination." This is an extremely critical oversight. The timing of conception is but one aspect of the Shettles technique. His further recommendations concerning artificial alteration of the vaginal environment and cervical milieu, deposition of sperm, and so forth (all described in detail later) *have the effect of mimicking the conditions of artificial insemination.* Dr. Shettles observed and published data on

the preponderance of male offspring resulting from artificial insemination years before Dr. Guerrero came to the
same conclusion; it was only natural that in developing his
sex-selection method for conceiving males, Dr. Shettles
should attempt to reproduce the conditions that prevail in
artificial insemination. So let us emphasize: the Shettles
procedure, though it does not really introduce anything
alien and certainly nothing harmful into the system,
is not natural; if it were, we would all still be stuck in
Mother Nature's "50–50 Club," taking what we *get* instead of what we *choose.*

The only other challenge to the method prescribed by
Dr. Shettles has come from William James, a nonmedical
British researcher. James has accepted on faith the Guerrero thesis that in natural conception, insemination close
to the time of ovulation will result in a girl and insemination several days before will result in a boy (the opposite
of what our data and experience have shown to be true).
Thus, he hypothesizes, a woman having sex every day or
nearly every day is more likely to have a boy. He attempts
to back this up with statistics, which he says show that
among first births there is an excess of boys to mothers
under age twenty-five, while women over thirty-five give
birth to more girls. The clincher here, he says, is "unpub-
lished data which powerfully suggest that coital rates [rates
of intercourse] in the first month of marriage are higher
than during any subsequent month."

We don't doubt that newlyweds *do* make love more
often, but we doubt that this results in more male off-

spring. (Other researchers have failed to find support for James' thesis relating frequency of intercourse and sex ratios; see for example Dwight T. Janerich in the medical journal, *The Lancet,* April 24, 1971.) As for the statistics indicating that women under twenty-five are more likely to give birth to male offspring (and please note that "more likely" does *not* mean "very likely") than are women over thirty-five, explanations other than frequency of intercourse appear far more logical. For one thing, the vaginal and cervical secretions of younger women are known to be more alkaline, more abundant, and in consistency and overall quality more conducive to the transport of sperm, a fact that should particularly favor the fleeter male-producing sperm. Moreover, the husbands of women under twenty-five are very likely to be younger than the husbands of women over thirty-five, and there is evidence that the ejaculate of younger men may contain greater numbers of sperm, and the sperm may have greater vitality, factors that in the Shettles theory favor male conceptions.

It is appropriate here to cite the findings of the Austrian researcher Dr. August J. von Boronsini, who found that potentates with harems had a very high frequency of intercourse and fathered an abnormally large number of female offspring. This is consonant with our observation that the man who has frequent intercourse diminishes his sperm count, a factor contributing to female offspring.

Finally there is James' surprising claim that Orthodox Jews in Israel have more female offspring than non-Jews. This he attributes to the Orthodox Jewish practice of *niddah*—abstinence from intercourse until a week after menstrual bleeding ceases, which would place the first intercourse during the cycle at or about ovulation time in most women. Data has existed for decades attesting to the surplus of *male* offspring among Orthodox Jews, as would be predicted by the Shettles theory. And since publication of our first book, Dr. Jacob Levy, a medical specialist in Jerusalem, has published a paper in the Hebrew-language *Koroth* (Volume 6, No. 5–6, November, 1973) reporting that the birth ratio among Orthodox Jews who are known to have *strictly* observed the *niddah* has been in the neighborhood of 100 females to every 130 males and "sometimes even higher." The average birth ratio for non-Orthodox Jews and most others is about 100 females for every 105 males.

Dr. Levy cites "the new theory of Shettles" as a logical explanation for the preponderance of male offspring among Jews who strictly observe *niddah.*

Dr. Levy writes,

According to this theory, there is a morphological difference between the two types of spermatozoa; the broad, robust sperm cell that carries the additional X-chromosome (which produces a female offspring) has more resistance to a damaging acid milieu, while the sperm cells that carry the Y-chromosome (which produces male offspring) are smaller and more agile in an alkaline surrounding. At the time of ovulation (about twelve days after the beginning of menstruation) the cervical mucus is generally alkaline, and therefore the Y-carrying spermatozoa will, at that time, often win the race to the egg cell to be impregnated and thus produce male children.

According to Jewish religious law, the ritual bath generally takes place twelve days after the beginning of menstruation. During this night, intercourse is a religious obligation. Offspring produced in this night will be mostly male. . . . The male surplus is a point in favor of the accuracy of Shettles' theory.

But remember that the theory has more to it than just the timing of intercourse. Abstinence until ovulation is critical in the quest for a boy, as Dr. Levy observes, but other factors can significantly further boost your chances of success.

Other partial confirmations of the Shettles theory have come from India, Germany, Poland, France, and England. The Indian zoologist Dr. B. C. Bhattacharya several years ago was intrigued by the fact that farmers in his country preferred to bring their cows in for artificial insemination at sundown rather than earlier in the day, claiming that these sunset inseminations resulted in more bull calves. Dr. Bhattacharya began studying this and found that it was true! Cattle inseminated later in the day *did* produce more male offspring which, in contrast to the situation on the Rummins farm in England, were considered the most desirable.

Looking for a scientific explanation, Dr. Bhattacharya finally concluded that the two types of sperm drifted to the bottom of storage containers at different rates with the female-producing variety reaching bottom first, so that by

the end of the day far more of the male-producing type
were at the top. Since each day the sperm was drawn off
from the bottom of the containers first, more males would
be conceived from the remaining upper portions late in
the day. And what logical reason could there be for this
segregation of the two types? A difference in weight, of
course, the same difference upon which Dr. Shettles has
predicated part of his method; the male-producing sperm
are lighter and tend to remain suspended in the upper
strata of the seminal fluid, while the girl-producing sperm
are heavier and sink faster.

Dr. Bhattacharya later moved on to the famed Max
Planck Institute for Animal Breeding at Hagen, West
Germany. There he undertook a series of experiments with
rabbits, inseminating thousands of them with sperm that
had settled under a variety of conditions. At first he got
mixed results, but he finally hit upon a formula that made
"sedimentation," as he called his technique, look promising. The trick was to refrigerate the sperm (mixed in a
protective solution of egg yolk and glycerol) for about
twelve hours at a temperature just above the freezing
point. This prevented the sperm from swimming about,
and in their relatively immobile state, they separated far
more readily than they had previously.

Dr. Bhattacharya used the refrigerated samples to im-
pregnate 176 rabbits. Those inseminated with sperm that
had settled to the bottom of the containers produced 72
percent females. The rabbits that were impregnated with
cells from the top of the samples gave birth to males 78
percent of the time.

Dr. Bhattacharya's findings have more recently been confirmed by two other European researchers, E. Schilling of the Max Planck Institute and P. Schmid of Zurich. They showed that the male-producing sperm of sheep are about 4.8 percent shorter and 6.7 percent narrower than the female type. They also demonstrated that the female-producing sperm of sheep had a significantly slower velocity than the male types—a difference of 25 to 28 percent! This tends to confirm Dr. Shettles' observations of *human* sperm.

In 1972, Dr. A. M. Roberts of Guy's Hospital Medical School in London provided further confirmation of this aspect of the Shettles theory—that male-producing sperm are not only lighter but also move much faster than female-producing sperm. Dr. Roberts did what Dr. Shettles had done several times: he placed sperm in tubes (the "physiological racetracks," as Dr. Shettles had called them) and watched to see which type would get through the fluid in the tubes first. Like Dr. Shettles, he found that the Y, male-producing sperm were most likely to win these races.

More recently still, a team of research scientists at Schering A.G., a Berlin pharmaceutical firm, consisting of Drs. R. J. Ericsson, C. N. Langevin and M. Nishino, have also confirmed the faster swimming ability of the lighter, more agile Y sperm. Dr. Ericsson has just relocated to the United States and has opened a laboratory in California, where he is developing his own technique of sex selection based on test-tube separation of the two types of sperm followed by artificial insemination. We will have more to say about this and similar developments in Chapter Nine.

The most direct new confirmations of the Shettles technique come from France and Poland, where procedures very much like those advocated by Dr. Shettles have been used with very nearly identical results. First came the report of Dr. Franciszek Benendo, published in *Polish Endocrinology* and subsequently reported upon in *Medical World News* (August 13, 1972). The article begins:

> About a year and a half ago, Dr. Landrum B. Shettles, at Columbia University College of Physicians and Surgeons, proposed a regimen for prospective parents who want to pre-determine the sex of their offspring. His prescription was based on such factors as the pH of the cervical and vaginal mucus at various times in the woman's menstrual cycle and controversial theories about differences in the size and shape of X- and Y-chromosome-bearing spermatozoa. He suggested that a major element in determining the conceptus' sex is the time of intercourse with respect to ovulation.
>
> Support for part of Dr. Shettles' thesis appears now in a study by a European physician. While he makes no comment on the presumed physiologic basis of the New York obstetrician's regimen, Dr. Franciszek Benendo of the County Hospital in Plonsk, Poland, has confirmed that the timing of coitus that leads to fertilization has a profound influence on the offspring of the sex.
>
> Dr. Benendo studied 322 couples in whom the date of fertilizing intercourse and the date of ovulation could be fixed. His first group consisted of 156 married couples "in whom the solitary sexual contact usually took place two to five days before the term of ovulation." In the second group were 18 couples who had intercourse two days prior to the woman's ovulation. And 148 couples who had coitus

in the period from one day preceding to two days following ovulation constituted his third group.

One hundred fifty-seven children—including one pair of
twins—were born to couples in the first group. Of these,
133 (84.7 per cent) were daughters and 24 were sons. Nine
children of each sex were born to the second group [in
which the fertilizing intercourse occurred two days before
ovulation]. And of the 151 children, including three pairs
of twins, born to the third group [in which the fertilizing
intercourse occurred at or near the time of ovulation], 131
were sons (86.8 per cent).

And the article concludes:

To explain his findings, Dr. Benendo introduces the con-
cept of biological potential. "It may be assumed," he
writes, "that the potential of the Y spermatozoa in the first
two days postinsemination [after intercourse] is higher
than that of the X spermatozoa. After two days, on the
average, the X spermatozoa begin to predominate, and
their potential becomes greater than that of the Y spermatozoa." According to his statistics, he notes, the magnitude
of the difference of the potential is approximately six to
one. Finally, "when coitus takes place on the second day
before ovulation, as a result of the potential of the spermatozoa and ova at the time of fertilization, 50 per cent of
the offspring will be male and 50 per cent female."

He adds that 11 of the couples included in his study had
come to him before the conception with the desire to have
children of a given sex. After they followed his suggestions,
"the results were 100 per cent concordant with the wishes
of the parents."

Dr. Shettles has been in personal touch with Dr. Ben-endo since the latter published his first results, and the two doctors continue to get very similar results with the techniques that they recommend.

Still more recent confirmation comes from France. Writing in the *Journal of Obstetrics, Gynecology and Biology of Reproduction* in 1975, researcher and physician Dr. B. Seguy of Nice noted that preselecting a baby's sex can be very useful in preventing sex-linked birth defects (more on this in concluding chapters) and observed that he had been able to achieve a nearly 80 percent sex-selection success rate using a method similar to Dr. Shettles' method.

What makes Dr. Seguy's findings especially fascinating is that the 100 couples he worked with and reported on in this study were previously *infertile*. They were a selected group: the husbands were known to have adequate sperm count and motility; the women, however, had highly irregular cycles and in some cases did not ovulate at all, prior to treatment. Ovulation was stimulated and their cycles regularized and made predictable by the administration of hormones called human gonadotropins. Very careful temperature records were kept, under the supervision of Dr. Seguy and colleagues, so that in most cases the precise time of ovulation was authoritatively confirmed. Dr. Seguy followed all of the Shettles recommendations except those related to vaginal douche. He endorsed the principle of the douches, however, and now asserts that his success rate would in all likelihood have been even higher had he employed the douches. Even so, by timing intercourse in each case at or near the moment of ovulation and following the other recommendations for coital positioning and so forth, 77 of the previously infertile couples gave birth to boys!

Finally, Dr. Shettles himself has made some new findings that powerfully support his earlier observations. In the course of recent infertility research, Dr. Shettles devised a technique for obtaining samples of sperm as they progressed through the female reproductive tract at various times after intercourse. In all cases, intercourse had been made to coincide closely with the laboratory-confirmed time of ovulation, so that the female environment was most conducive to fertilization and, according to the Shettles theory, to fertilization by male-producing sperm. Dr. Shettles was able to aspirate into a specially devised catheter (harmlessly introduced into the vagina, cervix, and uterus) secretions containing sperm traveling toward the egg in the Fallopian tube. The farther along in this journey that he sampled, the greater the proportion of male-producing sperm he encountered. That the male type were preponderant was confirmed not only by phase-contrast examination but also by a relatively new technique employing a fluorochrome quinacrine dye that "lights up" the Y chromosome (but has no effect on the female X) when viewed through a fluorescence microscope. A "lit up" or fluorescing spot is regarded by virtually all researchers in this field as *proof positive* of a Y-carrying, male-producing sperm.

CHAPTER 6

Boy: Easier?

Let us say at the outset that the procedure for begetting male offspring is somewhat simpler and generally less time-consuming than the procedure for female offspring. The instructions in this chapter, however, must be very closely observed. Not all women or all men are alike; differences must be taken into account and adjusted for.

In preceding chapters we have documented the relationship among timing of intercourse, ovulation, and sex of offspring. We have shown that if a couple has intercourse as close as possible to the time of ovulation, they are more likely to conceive a male child, especially if they follow certain other recommendations as well. The fleeter, more agile male-producing sperm find the conditions within the female reproductive tract most favorable at or near the time of ovulation, and they race ahead of the more cumbersome female-producing sperm. If there has been a miscalculation in timing, however, and an egg is not waiting (that is, it has not yet emerged from the ovary where it develops), then the chances grow greater with each hour of delay that the slower but more enduring female-producing sperm will catch up with their shorter-lived brothers and still be fit for the fertilizing task once the egg does arrive. For the boy, there can be no dawdling; you must be right on target.

Luckily for those who want sons, it is easier to come in on target than to play the "waiting game" for the girl as described in the next chapter. By the way, even if you're only interested in male offspring we strongly urge you to read the next chapter as carefully as you read this one in order to better understand what you must avoid as well as what you must practice.

In our first book we recommended that women use a "fertility test kit" as a partial means of discovering their time of ovulation. At that time there were some good test kits on the market, but there were also some bad ones—with the consequence that none of these kits is any longer available to the general public. Another aid we recommended was called Tes-Tape, which is still available without prescription in larger drugstores. Designed to help diabetics test the amount of glucose in their urine, Tes-Tape is a roll of special yellow paper that comes in a Scotch Tape-type dispenser. The tape turns varying shades of blue and green, depending on the percentage of glucose. And in the 1950s, Dr. Shettles and coworkers discovered that glucose is also present in cervical mucus and that it becomes *increasingly* abundant as ovulation approaches.

Though we do not recommend Tes-Tape as a primary means of finding the time of ovulation, we regard it as an

optional or additional aid that may be very helpful to
some women. Other women, owing to their particular body
chemistries, may find it difficult to work with and interpret.
For those who wish to try it, however, we offer some suggestions: each morning, beginning on the day after menstrual bleeding ceases in each monthly cycle, tear off a
three-inch strip of tape. Bend the strip over your index
finger (you'll have to sacrifice that fingernail temporarily)
and secure it to the finger with a small rubber band. Now
guide the finger into the vagina far enough so that the tip
of your finger and the tape make contact with the cervix.
Try to go as directly to the target as possible so that you
don't totally dampen the paper with the more acidic secretions of the vagina. How do you know when you are touch-
ing the cervix? Dr. Shettles says that it feels something like
the tip of your nose. Practice a few times before trying it
with the tape.

Hold the finger gently up against the cervix for ten to
fifteen seconds, then withdraw it directly and quickly. Note
the color of the tape at the tip of your finger, where you
made contact with the cervix. Early in the cycle it probably won't change color at all. Or it might change to a
light green. As you approach ovulation, each new tape
should be darker and darker where it has made contact
with the cervical mucus (which, you recall, becomes increasingly alkaline as ovulation nears). To determine when
ovulation is about to take place, consult the color chart
on the Tes-Tape dispenser; when the color is close to that

of the darkest color on the chart (a deep greenish blue), ovulation is most likely at hand.

You will notice that the Tes-Tape chart is coded for the urine test, but it can also be used for detecting ovulation, as described above. However, individual women differ in their responses to the tape, so that if it is used at all *it should only be used in conjunction with the other methods that we are about to describe.* The tape *can* serve as a useful aid in determining time of ovulation. But if the results vary a great deal or seem generally confusing, then you shouldn't place much confidence in the tape. The only way to find out how well it works for you is to experiment with it through a few cycles—before using it in any attempt to choose sex. Then you can see if the color is darkest on the same day of your menstrual cycle each month. If you have a lot of time, you might want to use it for six months or so; then if you recorded the darkest color on the same day for four out of six months and came within a day on the other two months, you could be sure the method was working pretty well for you. Keep careful records of the color you come up with on each day that you try the Tes-Tape, taking special care to mark the day of each cycle on which it turns darkest.

One reason many women found the Tes-Tape less than ideal was that it would turn very dark even early in their cycles, which they couldn't understand. Fortunately, other signs made it clear that ovulation was not yet at hand. We have discovered now that in most instances the darkness of the tape was caused by their husband's seminal fluid; if these women had intercourse the night before they sampled the cervical fluid, the tape would pick up the highly alkaline seminal fluid still in their reproductive tract and record that.

Because of this difficulty, we do not recommend that Tes-Tape be used when a couple desires *female* offspring. The reason for this is that the procedure for female offspring calls for frequent intercourse, nightly if possible, up to a certain number of days away from ovulation, depending upon individual circumstances. But in the method prescribed for *male* offspring, *total abstinence from sexual intercourse—from the beginning of the cycle up to the time of ovulation—is absolutely necessary.* Thus, if our method for male offspring is followed faithfully, there will be no seminal fluid in the vagina and around the cervix to interfere with the Tes-Tape, so we recommend using it in conjunction with other techniques when male offspring are sought.

To continue with the procedure for a boy: Use the Tes-Tape first thing every morning, before rising or eating. No sexual relations are permitted from the start of menstrual bleeding until the time of ovulation. Some couples have asked whether they might engage in intercourse if the husband withdraws before he reaches orgasm. We have to say no for two reasons: 1) small amounts of seminal fluid may still enter the vagina, enough to influence the outcome of the Tes-Tape check for ovulation and, more important, 2) that same small amount of seminal fluid (which many men secrete prior to full ejaculation) may contain viable sperm, which are capable of fertilizing the egg. Nor, for reasons soon to be explained, should the husband use a condom or seek sexual release through masturbation. *No sex whatever for either husband or wife is the rule during the early part of the cycle.* (That's why Dr. Shettles says, "It's more fun to have little girls.") As ovulation approaches, the woman may benefit from frequent testing with the tape at different times of day; over a period of several menstrual cycles she may thus pinpoint not only the day but the approximate time of day when she seems most likely to ovulate.

A certain number of lucky women are able to rely upon something called *Mittelschmerz* to pin down the time of ovulation. Many doctors regard *Mittelschmerz* as the best natural indicator of all. For those not familiar with this harmless, and in our case very helpful phenomenon, it is a pain that is felt in the lower abdomen, often on the right side, at the time of ovulation. Some women feel a sharp twinge of pain at the precise moment that the egg bursts out of the ovarian follicle. Some experience a small amount of bleeding at the same time. If you have such midcycle pains or bleeding, check with your doctor; it could be the telltale *Mittelschmerz.*

About 15 percent of all women have these pains, but in other women the pains are nonexistent or vague. Some women who have *Mittelschmerz* have been operated on for what the doctors thought was acute appendicitis—only to go on experiencing the pains month after month. The late Dr. Sophia Kleegman, another sex-selection pioneer, always alerted her patients to these pains and, in fact, taught 35 percent of them to become aware of the pains by practicing what she called the "bounce test." From the ninth day of the cycle, both morning and night, the patient was instructed to bounce on a hard surface such as an unpadded wooden chair by sitting down abruptly three or four times. If the woman felt the pain she was told to note the day and repeat the test during the next cycle to see if a pattern emerged. In this procedure the pain experienced upon bouncing might not signal the exact time of ovulation, but it could reveal that the egg had emerged recently or that it would soon burst out of its follicle. In either case, there is still time to conceive a boy.

While not all women can benefit from *Mittelschmerz* or the use of the Tes-Tape, almost all can make valuable use of what we call the "stretch test." Again, we do not consider this method sufficient unto itself, but it can provide valuable clues. Let us explain. The cervical mucus is generally thick and rather milky in appearance, but it becomes increasingly thin and clear as ovulation approaches. At ovulation it has the consistency and transparency of raw egg white. At that time the mucus can be stretched easily and may be so abundant that spotting of underclothing occurs. By careful observation of the consistency of the cervical mucus, from sticky to most fluid, some women can come close to finding their day of ovulation; this is especially useful in conjunction with Tes-Tape.

For most women, however, we now believe that the best method of determining time of ovulation is by measuring basal body temperature (BBT). Considerable experience with BBT in the years since publication of our first book has convinced us that most women can adequately master the BBT technique and that, especially when used with the other methods described above, it offers the best hope of coming in on target. The BBT method is based upon a well-established sudden shift in body temperature that occurs at or near the time of ovulation.

The reason for this temperature shift is as follows: during the early part of the menstrual cycle, the egg undergoes final maturation in the ovary under the influence of the hormone estrogen. During the time that this powerful hormone prevails, it has the side effect of holding down body temperature slightly. When the egg erupts from the follicle on the surface of the ovary, however, an equally potent hormone, progesterone, takes over, in part to nourish the lining of the uterus, readying it to receive the fertilized egg. One effect of progesterone is to abruptly raise body temperature and keep it elevated throughout pregnancy, if that should occur. If the woman does not become pregnant, the progesterone will "switch off," and her body temperature will fall back as the next menstrual cycle approaches.

So the task is to detect this upward temperature shift and take advantage of it. This is commonly done by keeping a BBT chart (see Figure 5) on which you record your body temperature daily. You can buy graph paper or rule off plain paper to make charts modeled after Figure 5. Put your name and date at the top of the chart as indicated. Your cycle begins (Day 1) with the first day of bleeding. Your bleeding should be indicated by blacking out the 98.0° squares, as shown. It is not necessary to take your temperature during bleeding days, since 98.0° is used as a basic reference for *all* women. Duration of bleeding differs from woman to woman; it often lasts about five days, sometimes as little as three days or as long as a week. When bleeding ceases, record daily temperatures by placing dots in the appropriate squares. Connect the dots with straight lines to show temperature patterns. (We will return to Figure 5 shortly.)

To take the basal body temperature you need a special thermometer; your doctor can recommend one, or you can obtain one through your pharmacy without prescription. Ask for a BBT or ovulation thermometer; they are the same. Most have calibrations from 96 or 97 degrees to 99 or 100 degrees, with easy-to-read markings for each one-tenth of a degree in between.

Remember that we're trying to measure *basal* body temperature, that is, your body's *base* temperature, influenced as little as possible by activity, stress, and other factors that raise temperature above the base line. Thus it *must* be taken first thing in the morning—even before you get out of bed. You must not eat, smoke, drink, indulge in sexual play or intercourse or other activity between waking and taking the temperature.

Take your temperature orally. (Be sure to follow all the manufacturer's directions on handling, storage, and care of the thermometers.) Place the thermometer under your

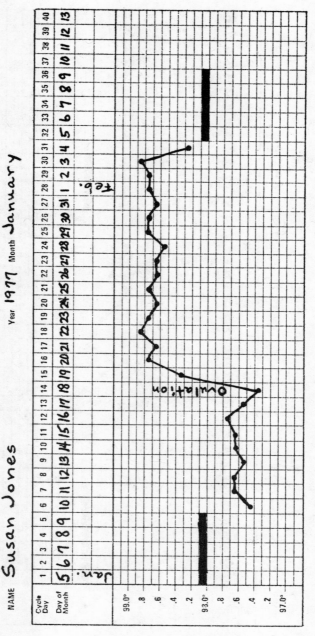

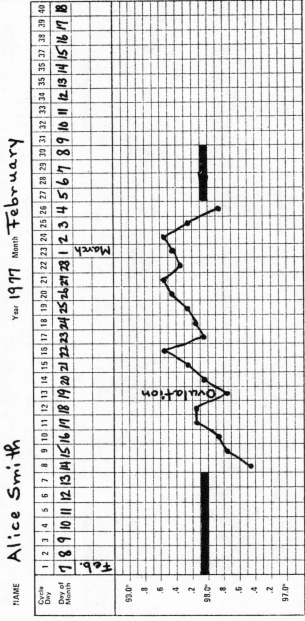

BBT CHART

NAME Alice Smith   Year 1977 Month February

BBT CHART

NAME Judy Klein   Year 1977 Month March

BBT CHART

NAME Maria Ramez

Year 1977 Month June

tongue for exactly three minutes. Be precise in your timing, watching the sweep second hand of a watch or clock.
Keep your hands off the thermometer while it is in your
mouth.

At the end of the three minutes, note the temperature
carefully and record it on the chart by placing a dot in the
appropriate square. Do not confuse the days of the calendar
month with the days of your menstrual cycle. Repeat this
procedure each morning, and try to take the temperature
at about the same time each day. And during the period
in which you are trying to assess your ovulatory pattern,
try to maintain a fairly regular schedule of activity, diet,
sleep, and so forth, to ensure that you're getting the best
possible readings.

If you have a cold, fever, or upset stomach or if you are
under strong emotional stress, sleeping poorly, keeping
late hours and getting only a few hours sleep, smoking
heavily, or drinking a lot of alcohol, your readings will be
less stable and not as reliable. But even if you're hit by a
cold or your regular schedule is otherwise interrupted, just
keep taking your temperature until you feel you have
recorded enough stable days and enough cycles in which
you were relatively free of such stresses to make a valid
assessment of your ovulatory pattern.

Even though some of these difficulties will inevitably
crop up, over a long enough period of time (and in some
cases in a relatively short time) a clear pattern will
emerge. Studies of vast numbers of women show that far
more are regular in their menstrual cycles than are irregu-

lar. Many women average a twenty-eight-day cycle, though there will always be small variations. Some find that their cycles are always longer than that—up to forty or more days. Others have regular but very short cycles, perhaps only twenty days long. Some women, unfortunately, wander all over the place, one month having a twenty-four-day cycle and the next a forty-day cycle. They must take special care in their effort to determine time of ovulation.

Continue to record your temperature on the chart each day of your cycle, which ends when you begin to menstruate again. Start a new chart on that day. What you must watch for is the sudden shift in temperature that signals ovulation. Temperature on the days before ovulation is always noticeably lower than on the days after ovulation. *Before* ovulation the temperature is *usually* in a range between 97.4 and 97.8 degrees. *After* ovulation the temperature increases, *usually* by four-tenths to one full degree in a single day. Thereafter until the end of the cycle, even though the temperature often dips up and down it will almost always remain above 98 degrees, often lingering between 98.2 and 98.6 degrees. As bleeding approaches it will drop down sharply. But again there may be variations from these averages because of individual body chemistry or such factors as smoking, late hours, illness, etc. If the chart seems erratic, examine your recent activities, state of your health, and so forth. You may find it useful to keep a diary along with your charts, noting in detail how you felt each day, whether you had taken medication and, if so, what type. You can also use this diary to keep track of dates of intercourse, dates of abstinence from intercourse, and anything else pertinent to the effort to beget a boy.

Let's return now to Figure 5. As you can see, menstruation ended on day 5. On day 6, the woman entered her first temperature recording—97.4 degrees (note that the proper box for each temperature is *above,* not below, the line indicated). Now look at cycle days 7 through 12: the temperature bounces up and down a bit but still remains low throughout this period. On day 13 the temperature has gone down a little and by day 14 it has dipped further still. Then on day 15, the temperature has zoomed upward a full degree. This marked upward shift indicates that ovulation has taken place, as further confirmed by the fact that in the next several days the temperature remains high, not dropping significantly until the next menstrual bleeding begins, starting a new cycle. (Remember to keep a separate chart for each menstrual cycle.)

Studies of the menstrual cycles of thousands of women have shown that the pattern depicted in Figure 5 is typical; these studies have shown that there is often a fairly marked dip in temperature, as seen in Figure 5, just before the sudden, even more obvious rise in temperature. There is still controversy over whether ovulation takes place at or about the last low temperature before the marked rise or on the day of the marked rise itself. Probably it takes place between those two events—in other words, in the case illustrated by Figure 5, between the morning of day 14 and the morning of day 15.

When trying for a boy, our experience shows that you

should time intercourse immediately *after the rise* is noted; for example, again using the case illustrated by Figure 5, on the morning of day 15. On the morning of day 14, this woman *might* be fairly certain, if she had become thoroughly familiar with her ovulatory pattern over a period of months, that the rise in temperature would almost certainly occur the next morning, but if she waits she will be absolutely certain. And even if ovulation occurred on day 14 and she waited until day 15 to have intercourse, our data show that her chances of conceiving a boy would remain excellent.

If you're confident that you've located the day on which the temperature is at its last low point before the upward shift, you might want to schedule intercourse for that evening, hitting it in the middle, so to speak. But for most couples we recommend waiting until the morning when the upward shift is actually noted. And remember to look for other signs of ovulation as you practice over three or more months—consistency of cervical mucus, *Mittelschmerz* if any, color changes in the Tes-Tape.

Now let's look at a couple of charts depicting temperatures that do not follow such a typical and easily interpreted pattern as that seen in Figure 5. In Figure 6, note that menstrual bleeding is again marked with blacked-out squares. Bleeding ends on day 7. Temperature starts at 97.4 and rises to 98.1, where it lingers for two days before abruptly dropping to 97.7 on day 13. On day 14 it rises again, but only three-tenths of a degree to 98 degrees even. Now if this woman were taking her chart for the first time or if she knew she had a highly irregular cycle, that three-tenths rise might not be enough to make her feel confident that the basic upward shift had taken place and that ovulation had occurred. She might worry that on the next day her temperature might dip again. As we can see, however, it did not. Should she wait, you may ask and, when she finds that the temperature does continue to rise, have intercourse the next day? *No,* because in all probability, ovulation has occurred on day 13 or 14. To wait until the fifteenth day of the cycle would put her in jeopardy of conceiving a girl. Within one day after ovulation the chances are still better for a boy. But two days after ovulation, if conception occurs at all it might go either way, according to our data. Only by practicing for several cycles can you become familiar with the way your body works. Generally the pattern will be more clear-cut than this.

But occasionally the situation may be even more confusing, as in the chart in Figure 7, which is also based on a real case. Note that there is a marked drop between days 12 and 13, followed by a sharp rise on day 14 and an even sharper drop on day 15. There is a moderate rise on day 16, a further small rise on day 17, a small dip on day 18, and a very substantial rise on day 19. Which of the three dips most likely signaled that ovulation was at hand—the one on day 13, day 15, or day 18? This woman was able to determine that ovulation occurred on day 15 by studying other signs and by familiarizing herself with her cycle over a period of months. On most months she ovulated on day 15, occasionally on day 14, as indicated by charts that were generally easier to understand. If she had been less experienced she might have decided that ovulation was taking place on day 13, for there was a low followed by a rise of half a degree. But she knew she had never ovulated that early before, so she decided to wait and see if the temperature would fall back again, which it did. As for the dip on day 18, that was just too late, given her pattern—and even if that were the right day she wouldn't have risked the attempt with so many doubts in her mind. She would have waited until a more familiar cycle asserted itself.

When trying for a boy, do not select a cycle in which your temperature readings are ambiguous or out of whack with your past cycles. Wait for a very clear-cut temperature shift, and then don't try the *first* time you get such a shift. It could be a false alarm caused by some external stress factor, as in the case shown in Figure 7. You've got to familiarize yourself with your pattern; we can't emphasize that enough. If you're worried about ups and downs like those seen in Figure 7, you might try initially looking for peaks that in one day zoom up above 98 degrees and preferably above 98.2 degrees before assuming that the upward shift has actually taken place.

In a few cases the upward shift may creep rather than zoom. In the chart in Figure 8, ovulation could conceivably be occurring between day 11 and day 12, or it could be occurring between day 12 and day 13. Generally, a woman noting the small rise between days 11 and 12 would not believe that ovulation had occurred, thinking the temperature might dip again. But noting the greater rise between days 12 and 13, she would wonder if she had missed the ovulation date. A case like this is difficult to interpret. If month after month the same pattern emerges, then the woman trying for a boy would probably be best advised to consider day 12 as the critical rise, even though it is a small one.

As the day upon which you attempt to conceive approaches, you must keep in mind a number of facts. You must abstain from intercourse from the beginning of the cycle until the target day so that a girl is not conceived accidentally and so that the husband's sperm count is at its maximum. Our data indicate that the greater the sperm count the greater the chances of conceiving a boy.

After the target date you may resume intercourse, but the husband should use a condom. Contraceptive foams and jellies should not be used by women who are hoping to conceive a boy because these products are usually acidic and therefore detrimental to male-producing sperm.

Women who have been on the pill should not attempt to conceive for four to six months after discontinuing it. Studies show that women who conceive immediately or within a few months after going off the pill are more likely than other women to suffer miscarriages and other complications of pregnancy. Moreover, it takes time for your menstrual cycle to get back to normal. During this period you can keep temperature charts until it is evident that your normal cycle has reasserted itself. The IUD does not adversely affect the cycles of most women, and testing can proceed while you have the device in place. Still, it is a good idea to wait a few months after removal of the device before seriously attempting to conceive. No adverse effects are known, but it's better to be on the safe side.

When trying for a boy, the husband can contribute further to the cause by avoiding tight-fitting clothing for several months, particularly snug jockey shorts or jockstraps. It has been shown that if worn consistently, tight garments definitely and significantly reduce the sperm count. And a reduced sperm count, remember, increases the chances of begetting female offspring. It has even been demonstrated that tight-fitting clothes can result in temporary male sterility. The reason for this is that these garments trap body heat, and the higher temperature is sufficient to kill sperm of all types—but the smaller, more delicate male-producing type are killed first and in the greatest numbers.

Heat is such an effective sperm-killer that a team of medical researchers at the University of Missouri Medical School is now developing a contraceptive technique based upon exposure of the male testes to various heat sources. Other researchers have reversed some cases of male sterility by advising their patients to switch to looser-fitting clothing and to avoid high temperatures in their jobs. Some men with very low sperm counts have increased those counts significantly by lowering their testicular temperatures via frequent cold baths and even by immersing their testes in crushed ice—a technique that one medical wag calls "putting the rocks on the rocks." We aren't recommending the ice for men who want sons unless their physicians advise it, but we are recommending that you avoid tight pants, jockey shorts, and jockstraps.

While discussing the husband's role in sex selection, we should observe that a *very few* men produce sperm of predominantly or only one type. Even if you have two or three or more girls, however, you don't necessarily fall into this category; chances are that you do not, that you have been the victim of bad luck rather than of girl-producing genes passed on by your parents. But if a man's brothers have also fathered mostly girls, then an inherited trait looms as a greater possibility. Even so, using these procedures will give you a good chance of tipping the balance in favor of the outnumbered male-producing sperm. (See some of the letters in Chapter One.) The number of men who produce *exclusively* one type of sperm are so few as to hardly warrant having your sperm analyzed before trying for a child of the desired sex. Sperm analysis is done at only a very few centers and is scarcely worth the bother and expense.

There is one other thing the son-seeking husband can do, besides depositing the sperm in the proper way and at the proper time. He can drink two or three cups of strong coffee fifteen to thirty minutes before having intercourse with his wife on the suspected day of ovulation. Not for courage, but in order to increase the speed and staying power of his male-producing sperm. A number of researchers have recently published papers documenting the fact that caffeine has a stimulating effect on sperm. The caffeine will benefit both types of sperm, but it will benefit the androsperm the most, Dr. Shettles believes.

One group of researchers has suggested that male infertility caused by lack of sperm vitality may be overcome if the wives of such men use a coffee douche prior to intercourse; the caffeine in the vagina should give the sperm a boost. But Dr. Shettles has found that if the male takes the coffee "systemically," that is, drinks it, the effect is as good or better. Even the sperm of normally fertile men get an extra boost. This is an optional recommendation; some people dislike coffee or object to it on religious grounds. It isn't necessary but it can help.

Returning now to the wife and her preparation: if she knows she has highly acidic vaginal and cervical secretions, if she has an acidic stomach condition or ulcers or is under constant stress that might contribute to greater acidity, she can benefit from the systemic intake of small amounts of potassium iodide. How can you tell if you are more acidic than normal? If you have ulcers or an acid stomach you probably are all too aware of it. But beyond that, pay attention to your cervical mucus. If it is very thick, in poor supply, or if it does not become significantly clearer and more easily stretched as ovulation approaches, then it is likely to be acidic. Dr. Shettles and others have observed that potassium iodide can dramatically change the quality of the cervical secretions, making them more abundant, less thick, and less resistant to sperm migration. Writing in the journal *Drug Therapy* (Vol. 6, No. 5, May, 1976, pp. 191–192), Dr. Shettles reported that several women who had been infertile because their secretions were so highly acidic and hostile to all sperm had successfully conceived after taking potassium iodine for five days.

Dr. Shettles recommends the use of potassium iodide (ask the pharmacist for the nonprescription *saturated solution,* which often comes in a 15-cc bottle with dropper) in the amount of ten drops daily for five days prior to ovulation. The solution should be taken in a glass of orange juice. Potassium iodide has been used medicinally for decades; it is not known to have any untoward side effects, although a few women find that it temporarily increases salivation, and a very few women may develop a skin rash. Dr. Shettles recommends using potassium iodide in cases where secretions are insufficient or hostile and in cases where two or more girls have been conceived already and a boy is desired. It may also be used during any attempt at conceiving male offspring. It's a little more trouble but, in many cases, well worth it.

Many people have asked us about general diet as a possible influence on the outcome of the sex-selection effort. We believe that an individual who has a balanced diet will encounter no special difficulty, but it is obvious, in terms of the findings on potassium iodide and caffeine, that what we eat *can* have significant impact on even the subtlest body processes. In this connection, we should point out that there have been isolated reports that more males are born to parents who drink hard water than to those who drink soft water. Hard water is more alkaline in content.

It is possible that a few people, through some strange quirk or craving, have tended to eat foods that are mostly alkaline or mostly acidic and that they have thus altered their pH (acid/alkaline) milieu to the extent that they are more likely to conceive one sex than the other. In the appendix to this book we give a list of acid and alkaline foods, which you ought to at least glance at to ensure that you aren't too far out of balance. In general, even if you're trying to conceive a girl, you should be eating far more of the alkaline than the acidic foods—simply because the stomach can't tolerate too much acid. See Appendix A.

Whatever else you do to enhance alkalinity, it is essential that before intercourse on the day of ovulation, the woman should use a baking-soda douche, consisting of two tablespoons of baking soda to a quart of lukewarm water. Let the solution stand for fifteen minutes before use so that the soda dissolves completely. (This douche has been used for decades for a variety of reasons without ill effects of any kind.) Intercourse should follow the douche immediately. And in order to mimic artificial insemination in some respects, Dr. Shettles recommends vaginal penetration by the male from the rear. This, he explains, helps ensure that the sperm is deposited at the entrance of the womb rather than in the two spaces adjacent to it, the cul-de-sac and the posterior fornix. This is desirable because the secretions from the cervix and womb will be more highly alkaline than the vagina, even with the alkaline vaginal douche, and an alkaline environment is most favorable to the androsperm.

If the wife normally experiences orgasm, it is a good idea, if possible, to have her reach climax first or at the same time as her husband. This increases the flow of alkaline secretions, and her orgasmic contractions help transport the sperm into the cervix and beyond. *Multiple* female orgasms are even better. And *deep* penetration by the male during his orgasm will further ensure that the sperm are deposited as close to the cervical opening as possible.

If there is a long history of female offspring in the family and if for some reason (such as highly irregular cycles) you find it difficult to follow these instructions, as a final option you might want to consider artificial insemination with professional assistance in timing ovulation. If you do (of course utilizing the husband's sperm), remember to follow all of the above procedures anyway.

We believe that at least 85 percent of the couples who use our method will succeed if they conscientiously follow all the procedures in this chapter. *As soon as you have made your effort, please fill out the Reader Questionnaire at the end of this book (Appendix B), save it until your child is born, then tell us the final result in the space provided and mail the questionnaire to us.* This will help us immensely in establishing the success rate of our method and in making future refinements.

If you still have questions, you should be able to find some or all of the answers in Chapter Eight.

CHAPTER 7

Girl: More Fun?

"It takes more patience to conceive a girl, but it can be more fun," observes Dr. Shettles. The nearly two weeks of sexual abstinence necessary in the conception of a male is not required or recommended in the procedure for begetting daughters. But the timing for a girl is trickier than for a boy, and it often takes longer to successfully conceive.

We have shown that intercourse must cease two or three days before ovulation in order to maximize the chances of conceiving a girl. If the egg is already in the Fallopian tube or just about ready to descend into it at the time of intercourse, the smaller, faster male-producing sperm will get to the target first. Therefore we advise that intercourse take place two or three days in advance of ovulation, so that when the egg finally does arrive, only the sturdier, longer-lived gynosperm, which produce girls, will still be "in waiting" and capable of fertilizing the egg. The closer you come to ovulation the greater your chances of getting a boy. Thus you must begin cautiously, well ahead of ovulation, and, then, if you do not succeed in becoming pregnant at three days prior to ovulation during that cycle, the next time you should try at two and a half days and then, if still not successful, at two days. After that you leave the safe zone and risk begetting a boy.

It is absolutely essential that couples interested in conceiving a girl read the preceding chapter carefully, for there we describe some basic concepts and explain various tests that apply as well in this effort. (That is why the Girl chapter is shorter than the Boy chapter, lest we be accused of overemphasizing the latter.) In addition to the necessary background material, the Boy chapter provides information on what to avoid.

Again the first task is to determine the time of ovulation, so that intercourse can be scheduled to precede it by three days, initially. The Tes-Tape method discussed in the last chapter will not serve as well here as it does for those seeking a son. It cannot be used during the cycle in which the attempt at conception is made, because the recommended frequent intercourse (details later) for those wanting daughters would interfere with the outcome of the test. Seminal fluid is highly alkaline and remains in the vagina for some time; its presence would bring about color changes in the tape that would have nothing to do with the woman's cycle. (See Chapter Six for more details.)

The Tes-Tape may be used, however, during practice cycles in which the woman tries to determine her ovulatory pattern—provided the husband uses a condom to prevent any seminal fluid from entering the vagina and interfering with the test. If the woman uses contraceptive foams and jellies, which are usually acidic in base, these will also have an effect on the tape and thus make it unreliable. But if such foams and jellies are avoided and the husband always uses a condom, then the Tes-Tape can be useful to some women as an aid in determining time of ovulation.

At this point, some women may be asking, "Why all this fuss over foams and condoms? I'll just stay on the pill until I'm ready to conceive." That would be a mistake. In the first place, you need to allow four to six months after discontinuing the pill for your system to return to normal before attempting to become pregnant at all; in the second place, studies have shown that women who become pregnant soon after going off the pill run a greater risk of suffering miscarriages; and in the third place you need to use those four to six cycles to discover your time of ovulation.

If you wish to use the Tes-Tape during those preparatory cycles (*not* during the actual cycle in which you attempt to conceive), read the instructions in the previous chapter. We recommend the Tes-Tape only as an auxiliary method of finding your ovulation date. In the last chapter also read about *Mittelschmerz,* the distinctive midcycle pains that for about 15 percent of all women pin down rather precisely the time of ovulation, and about the "stretch test," which gauges the changing quality of the cervical mucus as ovulation approaches. These aids should be used in conjunction with basal body temperature

(BBT) charting, which we now believe is the most reliable method of determining approximate time of ovulation, easily mastered by most women.

The BBT method, as explained in detail in Chapter Six, is based upon the abrupt upward shift in body temperature that occurs at the time of ovulation. The shift is caused by the hormone that takes over after the egg erupts from the ovary and enters the Fallopian tube. During the first part of the cycle, even though the temperature may go up and down a bit, it remains noticeably lower than during the second half of the cycle, after ovulation has occurred. For the majority of women who have fairly regular cycles, this shift should occur on the same day each month. But the only way to be certain is to take your temperature faithfully through several cycles and record these temperatures on a BBT chart. (See instructions and Figure 5, preceding chapter.)

When trying for a boy, we recommend that intercourse be timed immediately after the rise in temperature is noted; for example, using Figure 5, on the morning of the day 15. *But for a girl* we recommend that you regard the last *low* temperature (day 14 in Figure 5) as the ovulation date and schedule your *last* intercourse three days previous to this—*on day 11.* This is an extra touch of caution in the beginning, helping to ensure that you don't get too close to ovulation and thus beget a boy.

As explained in the last chapter, temperature charts are not always quite so easy to interpret as the one shown in Figure 5. We suggest you go back and reread those pages about interpreting less typical charts (Figures 6, 7, and 8).

Since some women have fairly irregular cycles we have established *two formulas* to help you decide on which day of the cycle, in advance of ovulation, you should *cease* intercourse. Let's look at two real case histories by way of illustration:

Case 1. Linda writes:

> My periods are very irregular, and I have never ovulated sooner than day 19. Taking my temperature every morning before I get out of bed works fine. I always go from about 97.6 degrees to 98.4 when I ovulate, but one month I ovulate on day 19 and the next on day 25. My cycle runs anywhere from 28 days to sometimes 45. Is there any hope for me, or is this a losing battle? I thought maybe I'd have intercourse (with proper douche, etc.) on day 17 for five or six months and then move on to 18, 19, etc. until maybe I could get pregnant with a girl. I'm in no real hurry. I'd wait four or five years if I knew I could have a girl.

Our recommendations in this case are: *try two formulas and use the one that yields the earliest day.* The first formula is simply to look at the records of four to six cycles (preferably more if you have time and your cycle is very irregular), take the earliest date on which you ovulated during those cycles, and subtract 3 days. Using this formula, Linda would subtract 3 from 19 and come up with day 16. The second formula is to take the number of days in your *shortest* cycle and subtract 14 (inasmuch as 14 days often *but not always* elapse after ovulation before new menstrual bleeding begins) to get the earliest probable ovulation date in *any* month. Then subtract 3 days from that to put you at a time that favors female offspring. So here Linda would subtract 14 from 28, which equals 14. And from 14 she would subtract 3 to get 11.

Now it's true that Linda might very rarely have a cycle that runs as short as 28 days. But since she has no way of knowing in advance, given her great irregularity, and since she is determined to have a girl and has plenty of time to work at it, we definitely recommend that she start with a target date of day 11. In other words, she and her husband should have intercourse as frequently as possible from the beginning of the cycle to day 11. The last inter-course should occur on the morning of day 11; intercourse at that time is mandatory. Thereafter they must abstain from intercourse until they know they are in a safe period (at least two days after ovulation). But with Linda's ir-regularity, it is difficult to know when the safe period is. So when they resume intercourse they should use a condom or diaphragm with foam (here the acid base does no harm).

In any event, Linda should definitely *not start* with a target as late as day 17. If after three or four months she still has not conceived with day 11 as the target, then she could advance to day 12, wait another few months, move to day 13, 14, 15, and so on. Bear in mind that hers is a very unusual case and that such difficulties, fortunately, are not encountered by the average woman. Still, we believe that Linda and other women like her can succeed if they are patient.

Case 2. Mary is not nearly so irregular as Linda, but six months of temperature charting show that she may ovulate any time between day 11 and day 15, a fair spread over so short a period of time. Her earliest ovulation time is day 11, and her shortest cycle is 26 days. So again let us apply the two formulas and see which day she should use as the initial cut-off point for intercourse. If we subtract 14 days from 26, we get 12; then if we subtract 3 days from that we get 9. This is the formula that yielded the earliest day in Linda's case. But let's see what happens when we apply the other formula to Mary's case. If we subtract 3 from her earliest ovulation date, which is day 11, we have 8. So this is the formula Mary should follow. She and her husband should have their last intercourse on the morning of day 8, and it would be helpful if they also had intercourse the evening of day 7. If Mary fails to become pregnant after following this schedule for a few months, she and her husband can advance the date of last intercourse by half a day (to the evening of day 8) and then wait several more months before advancing the cut-off date to the morning of day 9. Patience *will* pay off.

Now to repeat a few essentials: once you've determined the earliest cut-off date (three days before earliest possible ovulation), indulge freely in intercourse, without condom or any other form of birth control, up to that date. If you have intercourse after the cut-off date, be sure to use reliable birth control (condom or diaphragm and foam; condom would be best during the first few days after the cut-off date).

In the preceding chapter we discussed a number of ways that the husband could enhance the chance of getting a boy. One of those was to wear loose-fitting underclothing, switch from jockey shorts to boxer shorts, and so on. It has been demonstrated that even a slight increase in temperature around and within the testes reduces sperm count. And we have found that reduced sperm count may increase chances of female conceptions. The heat slows down sperm production in general but it kills the more fragile male-producing sperm first. So the man who would father a daughter may choose to wear tight-fitting jockey shorts during those periods when he and his wife are attempting to conceive. He might even take a hot bath or two. Just don't overdo it, or sperm counts may drop so low, temporarily, that no pregnancy will occur. (Sperm production bounces right back once the heat's off.)

In the last chapter we also advised the husband to drink some strong coffee fifteen to thirty minutes before intercourse, because laboratory tests have shown that caffeine stimulates the sperm and speeds them up. And since the male-producing sperm are faster-moving ones, the caffeine should further enhance their running ability, to the greater disadvantage of the more cumbersome gynosperm. Therefore, those husbands who want daughters should *avoid coffee, tea, and other caffeine stimulants* in the period when unrestricted intercourse is recommended. The effects of alcohol on sperm behavior are unknown.

You may wish to consult the acid/alkaline food list (Appendix A) to make sure that you are not encouraging a

very marked alkaline milieu by a totally unconcious avoid-
ance of acidic foods. Please be aware, however, that it is
natural to have a *strong* preponderance of alkaline foods
in your diet. It would not be good for you to start loading
up with acidic foods; just make sure that you have a well-
balanced diet. Given the marked effect of a substance like
caffeine on sperm, we can't discount the importance of
diet, but at this point the effect on sperm of acid-based
foods remains largely unknown.

In general, however, it can be said that stress, whether
physiological or psychological, tends to favor female conception. A jittery, high-strung woman will tend to have
a more acidic milieu, which reduces the effectiveness of
any male-producing sperm that enter her body. Also, various long-term stresses applied to the husband will tend to
diminish his sperm count and thereby contribute to the
likelihood of female offspring.

Deep-sea divers, whose testes are subjected to increased
pressure (which may have a similar effect to that of increased heat), are known to father disproportionately large
numbers of girls. There have been reports that high-alti-
tude flyers are more likely to father girls, and recently a
researcher in Copenhagen revealed that anesthesiologists
are more likely to give birth to girls. One of Dr. Shettles'
anesthesiologist colleagues has confirmed this, observing
that an informal survey among a number of anesthesiologists revealed a ratio of nearly four girls for every boy! Dr.
Shettles hypothesizes that the noxious gases inadvertently
inhaled by anesthesiologists reduce their sperm counts in
general. And the first to be "knocked over" by the absorp-

tion into the blood of these gases are probably the less hardy androsperm.

Once you are ready to try for a girl, the woman should administer an acid-based vaginal douche consisting of two tablespoons of *white* vinegar to a quart of water. This makes the environment within the vagina temporarily more acidic, a factor that favors the female-producing sperm. Administer the douche without fail before *every* effort.

The face-to-face or so-called "missionary" position should be assumed during intercourse. This position makes it less likely that sperm will be deposited directly at the mouth of the cervix, where they might escape the acidic vaginal environment that partially separates the girls from the boys. If the wife normally has orgasm, she should try to avoid it. Female orgasm increases the flow of alkaline secretions, which tend to weaken the acid environment of the vagina. Shallow penetration by the male at the time of his orgasm is recommended. This forces the sperm to swim through the acidic fluids of the vagina before reaching the more alkaline refuge of the cervix and womb.

You will note that at the end of this book (Appendix B) there is a Reader Questionnaire. *When you have made your effort, we would appreciate it if you would carefully fill out the form, save it until after the birth of your child, fill in the final result (boy or girl), and mail the questionnaire to us at the address provided.* This will help us establish the rate of success of this method.

If you still have questions, we hope you will find the answers in the next chapter.

CHAPTER 8

Questions and Answers: Is It True What They Say about Jockey Shorts? Do Women Who Take LSD Have Only Girls?

The following is a series of questions culled from the thousands we have received over the years. Some of them have already been answered, but the answers bear repeating for extra emphasis. Chapter references at the end of some answers tell where to look for more information.

Q: Isn't there some absolutely surefire way of determining exactly when ovulation occurs? A friend of mine mentioned a new device that could do this, and since I have very irregular cycles I'd like to know more about it.

A: We, too, have heard of the "electronic vaginal thermometer" and even something called a "vagina voltage meter," which supposedly can take most of the guesswork out of timing ovulation. The "intravaginal telemeters" (reported on in *OBGYN News,* Dec. 15, 1975), however, are still experimental and, in any event, require sophisticated auxiliary equipment for proper interpretation. It is not likely that such devices will soon become available for use except in large medical centers.

However, one method that employs these new devices could eventually have some impact on the home front. The National Aeronautics and Space Administration has developed transistorized thermometers that can be placed in the vagina, with a remote antenna tucked under the mattress. Signals from the thermometer are picked up on a bedside cassette recorder and later decoded by equipment in the laboratory. It's conceivable that in years to come, women who have trouble pinpointing their time of ovulation may be able to rent one of these space-age devices. This method might also help make the rhythm method of birth control less of a gamble than it is now.

In the meantime, however, we recommend that you use the basal body temperature method, for we believe that even those women with very irregular cycles can succeed in conceiving the sex of their choice if they follow our instructions and are patient. (Chapters Six and Seven.)

Q: We're trying for a girl, and the Tes-Tape doesn't seem to be of much use to me. Almost every morning it turns very dark, even though I haven't ovulated. Is there something unusual about me?

A: Since you and your husband are probably having intercourse nightly (as we recommend if you're trying for a girl), the seminal fluid, which is very alkaline and which remains in the vagina for some time, is turning the tape dark and giving you a false reading. This is why we now recommend that the Tes-Tape not be used when trying to conceive female offspring. You may use it during cycles in which you are trying to determine your time of ovulation, preparatory to the actual attempt, but then your husband should use a condom so that his seminal fluid will not influence the test. And please note that Tes-Tape should be used only as an aid, not as the primary means of pinpointing ovulation. Temperature charting should be given first priority. (Chapters Six and Seven.)

Q: Is it true that most women ovulate on the fourteenth day of the cycle?

A: The fourteenth day is often cited as the most likely day, but only because it falls in the middle of the "average" twenty-eight-day cycle. Not all women have a twenty-eight-day cycle; and among those who do, not all ovulate on the fourteenth day. See Chapter Seven for the two formulas that will tell you the safest time for the attempt to conceive a girl; Chapter Six will explain why a boy is easier to "time." Irregular cycles seldom mean that there is anything wrong with you, for some perfectly healthy women have markedly fluctuating cycles. Some have cycles that are always much shorter than average, say twenty days, others have cycles that are consistently longer, say forty days.

Q: We want to try for a girl. We understand that we must stop having intercourse three days before my ovulation. When can we resume sexual relations?

A: Actually, there need be no break in relations, but if you have intercourse in the several days after the "cut-off date" (three days prior to ovulation) make certain that your husband uses a condom. That is the recommended form of birth control for that time. However, it's not a bad idea to abstain from sex entirely for three to five days after the cut-off date, because there's a possibility that sexual activity during this period could upset a conception in progress. The male's seminal fluids and the female's orgasm both increase alkalinity, a factor favoring male conceptions. When you do resume sexual activity, be sure to use a condom and/or diaphragm and foam. (Chapter Seven.)

Q: I've heard that both diaphragm and rhythm failures usually turn out to be girls. Is that true and if so, why?

A: This does appear to be true. With the rhythm method of birth control, the point is to avoid intercourse on the days

approaching, on, and immediately after ovulation. But if intercourse occurs too close to ovulation, the woman may conceive, and the resulting offspring is most likely to be female,
as you would expect, according to our data. We have found
that this is also true of diaphragm failures. Most women use
an acidic foam with the diaphragm to further ensure against
pregnancy. This is usually quite an effective contraceptive
combination, but occasionally some sperm will slip through.
Those that survive the acid are more likely to be the thicker-
skinned, generally tougher female-producing sperm. The acid
foam, Dr. Shettles has found, acts like "a chemical sieve, filtering out the male while occasionally letting the female
through."

Q: We read about another sex-selection theory, developed
by a South American doctor, that seems to be at odds with
yours. Any explanation?

A: Half of the theory you refer to coincides with our own,
and we believe that there is a good explanation for the difference in the other half. See Chapter Five for details.

Q: Have other doctors or researchers been successful with
your technique?

A: Yes. And several important research efforts have confirmed the validity of our theory. (Chapters One, Four, and
Five.)

Q: Is there any truth to the idea that the time of year when
a baby is conceived will have an effect on its sex?

A: Many theories have been devised over the decades and
centuries related to sex ratio and season, phase of the moon,
and so forth. These are discussed in detail in Chapter Two.

Q: I've read that women who take or have taken LSD are
far more likely to give birth to girls than to boys. Could this
be true?

A: We hadn't heard of this, but when we made some in-

quiries it turned out that what you suggested does appear to
be true. We are not entirely surprised, since we have shown
(Chapter Seven) that unusual physiological and psychological
stresses will, if anything, tip the balance in favor of female
offspring. The reason for this is not difficult to discern: the
tougher gynosperm can put up with a lot more than can their
more delicate brothers. For example, heat reduces the numbers
of male-producing sperm faster than the female-producing
sperm. On the LSD issue, the one available sex-ratio study
showed that *all* the women studied who had been exposed to
this hallucinogenic drug gave birth to girls. The study was
small, however, and the results have not yet been confirmed.
As far as we know, no inquiry was made into what we regard
as the highly relevant question of drug intake by their male
partners. Interestingly, two independent studies have recently
shown that schizophrenic women also give birth predominantly
to female offspring. (See Chapter Seven for more details on
factors that seem to work in favor of female offspring, such as
exposure to noxious gases, high altitudes, and so forth.)

Q: Can what you eat have an effect on the sex of your baby?

A: We cannot provide a definite answer to this question
yet. As discussed in Chapter Six, however, we do know that
such substances as caffeine and potassium iodide can have
effects that favor male offspring, in an incidental way. We have
also cited reports showing that those who drink highly alkaline
"hard" water are more likely to give birth to boys. Diet is
discussed in Chapter Seven as well, and Appendix A at the
end of this book gives an Acid/Alkaline Food Chart.

Q: As a man, I must object to your statement that the male
is solely responsible for the sex of the child. He contributes the
male and female-producing sperm, but isn't it also true that
conditions within the woman can favor one or the other type
of sperm?

A: Looked at in this way, it is a fact that women share in

the responsibility. The woman whose secretions tend to be highly acidic, for example, puts her husband's male-producing sperm at a real disadvantage and in this sense helps "select" females. Similarly, some women, using various bits of misinformation, may insist on a particular timing for coitus that consistently favors one sex or the other. One woman who has three girls and no boys wrote that she had been told that when trying to conceive, she should engage in the first intercourse exactly five days after menstruation. It is not possible to assess the day on which this woman ovulates on the basis of a single letter, but it does seem unlikely that it would be as early as day ten. It seems considerably more probable that she ovulates on day twelve—two days after intercourse. At that point only the more sturdy gynosperm, which produce females, would still be able to fertilize the egg.

In another case along these same lines, however, it was the man who appeared to be responsible for an abundance of girls. As a truck driver, he was home only every other week, a circumstance that for three years caused him to have intercourse with his wife during the earliest portion of his wife's fertile period. Last intercourse was never later than the end of the twelfth day. As it turned out, his wife ovulated on the fourteenth day and never before: hence three little girls.

Q: Can the husband's or wife's age have anything to do with the sex of their children?

A: The sperm count of some males declines with age and this, as we noted earlier, can result in more female offspring. Similarly, as a woman ages, the cyclical nature of her body chemistry slowly breaks down, and the quantity and quality of her cervical secretions (which tend to be alkaline) diminish over the years. This deterioration tends to favor female offspring. And it is true that older couples do produce more female offspring than do younger people. One study showed that women of about fifteen, twenty, thirty, and forty years of age had offspring with sex ratios of 130, 120, 112, and 91 males, respectively, for every 100 females. The older woman, however (if she wants male offspring), can overcome this disadvantage by using the alkaline douche and timing procedures outlined in Chapter Seven.

Q: How soon before intercourse should the douche be taken?

A: The douche should be used within a few minutes before intercourse. In the case of the baking-soda (or bicarbonate of soda) douche, the solution should be permitted to stand for fifteen minutes before use. This allows the soda to become completely dissolved. The vinegar—which should be the white variety—will mix with water immediately and can be used at once.

Q: Should the douche be applied with a syringe-type applicator?

A: Dr. Shettles recommends the hot-water-bottle applicator. Let the fluid flow under the force of gravity alone.

Q: We are a little confused about the timing. What happens if intercourse takes place a few hours *after* ovulation? Will a boy or a girl be the most likely result?

A: According to Dr. Shettles' findings, a boy will be most likely during the six to twelve hours immediately following ovulation, since the cervical secretions are still generally very abundant and highly alkaline during this period.

Q: How long do the sperm and egg live?

A: Sperm cells have been known to survive for up to a week inside the womb. This, however, is extremely rare. Generally, the female-producing sperm will survive no more than two or three days. Male-producing sperm cells usually die within twenty-four hours. The unfertilized egg itself often lives for only twelve hours, rarely more than twenty-four hours.

Q: We are trying for a girl and don't want to take any unnecessary chances. We have decided to time intercourse for three days prior to ovulation, but our doctor tells us that this may keep me from getting pregnant at all. Is this true?

A: Dr. Shettles notes that the further away from ovulation that one times intercourse, the more difficult it is to achieve pregnancy. But it is also true that when one does achieve pregnancy in these cases, the offspring is very likely to be a girl. Since the couple in question doesn't want to take "unnecessary chances," it is wise to time intercourse well ahead of ovulation. If, after three or four months, they have still not achieved pregnancy on this schedule, they should move to a two-and-a-half-day interval and then, if that also fails, to a two-day interval. At two days it is still more likely that female offspring will result (provided the other procedures are also followed). But the couple wanting a girl has nothing to lose by starting out with the more cautious three-day interval. Pregnancy can and does occur in a significant number of cases under these circumstances.

It might be a good idea to reiterate here some of the clinical results of these timing procedures. "With exposure to pregnancy 2 to 24 hours before ovulation," Dr. Kleegman reports, "the babies were predominantly male (78 percent). With exposure to pregnancy 36 or more hours before ovulation, the babies were predominantly female." (Most exposures were 48 or more hours before ovulation.) Dr. Kleegman also reported that women using the rhythm method of birth control who became pregnant (from intercourse between days 4 and 7 of the cycle) gave birth to female babies 80 percent of the time. In another study, Dr. Shettles reported that one group of 22 couples who wanted female offspring took up to six months to conceive by consistently timing intercourse two to three days before ovulation. "Of 22 offspring," he notes, "19 were girls. In a group of 26 women anxious to have boys, the first coitus occurred at the time of ovulation or within 12 hours thereafter. To these women, 23 boys were born."

Q: We have given birth to three boys and no girls. My husband would like to have his sperm analyzed to see whether he is able to produce female sperm before we make another attempt. Can you tell us whether there is a center in our locality that can do this?

A: There are only a few doctors in the United States equipped to carry out these analyses at the present time, and most of them are involved in research rather than private practice. And Dr. Shettles notes that sperm analysis is rarely worth the trouble and expense. His studies show that most individuals, including those who have children of mostly one sex, are capable of begetting the other sex—simply by following the home procedures outlined in this book. The fact that a couple has children all of one sex is far more likely to be a consequence of bad luck than of genetics. Only in cases in which the husband's brothers, father, grandfather, and other relatives have consistently produced children of the same sex is it highly likely that there is a genetic factor at work. And when a man fathers nothing but girls, this might be a result of neither bad luck nor genetics, but of a low sperm count, a condition that can be partially remedied by abstinence from intercourse, as suggested in Chapter Six.

Q: Can the douche harm the fetus in any way?

A: Dr. Shettles says there is no danger whatever to the mother or the fetus. All babies born using these techniques have been completely normal.

Q: I read in a newspaper that the acid kills the male sperm and the alkali kills the female sperm. Is this right?

A: Some newspaper reports on Dr. Shettles' findings were misleading. The alkaline douche favors both types of sperm, but lets the male-producing sperm use the one advantage they have over the gynosperm: speed. The acid slows down both types but affects the gynosperm least, because they are larger and have greater resistance.

Q: I would like to use Dr. Shettles' sex-selection techniques, but as a Catholic, I am wondering if they run contrary to Church doctrine?

A: The Catholic Church does not object to the procedures. (Chapter Ten.)

Q: Is there a way to separate the two types of sperm, so that we would have a 100 percent chance of success? I am willing to undergo artificial insemination if there is any way to do this.

A: A number of exciting separation techniques are in the works or have been proposed; some have already been used successfully in animal studies, and early trials with humans also show great promise. (Chapter Nine.)

Q: I'm presently on the birth-control pill. How long should I wait after going off it before trying to choose the sex of my next baby?

A: Most doctors say three to six months. You need to let your menstrual cycle become regular again before trying to become pregnant; studies have shown that women who become pregnant soon after discontinuing the pill are more likely to suffer miscarriages. (Chapters Six and Seven.)

Q: How long does it take the sperm to get to the egg, assuming an egg is already waiting in the Fallopian tube?

A: About an hour in most cases.

Q: What happened to the fertility test-kits? I can't seem to find any.

A: They are no longer on the market. Some worked well; others did not, with the result that all were withdrawn from the market. Basal body temperature charting, as explained in Chapters Six and Seven, should be used to determine time of ovulation.

Q: I have a tipped uterus. Should I follow a different method from most women?

A: A good question and, in fact, you and other women like

you *should* follow a modified procedure. The condition in which the uterus is tipped toward the spine instead of toward the stomach (technically known as "uterine retrodisplacement") used to be corrected by surgery quite commonly. The doctor opened the pelvic cavity and shortened the ligaments that support the uterus, pulling the organ back into a more normal position. Today these operations are rarely performed; a tipped uterus is of little or no clinical significance, and a woman with a tipped uterus is often not even aware that she has one. The condition is relatively rare, but if your doctor has told you that you have one, you can alter the recommended procedures as follows. For a girl, follow the method without any changes. But for a boy, as soon as intercourse is complete, lie on your stomach, place a pillow under your upper thighs, and remain quietly in this position for about fifteen minutes. This will help ensure that the sperm immediately flow toward the cervix rather than into the posterior fornix, the space behind the cervix.

Q: Isn't it possible to get the sex of the child you want every time, using the new technique called amniocentesis?

A: Yes, but the procedure you name is not without risk, and in order to employ it as an aid in selecting sex, you must be willing to undergo abortion if the amniocentesis reveals that you are carrying a baby of the *undesired* sex. We discuss this *post-conception* sex-selection method in Chapters Nine and Ten.

Q: I have mixed feelings about the ability to choose sex. On the one hand I see that it might help people limit the size of their families by getting the desired sex right at the start. But what if there is some fad that makes boys more popular than girls, or vice versa? We could end up with terrible imbalances, couldn't we?

A: The most reliable data indicate that such imbalances will not occur. We discuss the sociological considerations of sex selection in Chapter Ten.

Q: My wife claims to have read or heard some place that the type of shorts a man wears can affect his fertility. If she weren't a nurse I'd say she was crazy. Is there any truth in this idea and, if so, could it have anything to do with the sex of a man's children?

A: The wife's claims are accurate. Dr. John Rock of Harvard, one of the developers of the birth-control pill, states: "Any clothing that prevents maintenance of an intrascrotal temperature that is at least one degree centigrade below body temperature will significantly lower sperm output. Daily wear of a well-fitting, closely knit jockstrap results in infertility after four weeks. . . . Normal output gradually is resumed after another three weeks without such interference. Enclosing the scrotum in ice for one-half hour daily may increase sperm output in perhaps 10 percent of moderately oligospermic (low-sperm-count) men and result in a long-awaited pregnancy."

Dr. Shettles, commenting on this, notes again that low sperm count is associated with a predominance of female offspring. So it would appear that a man can increase his chances of producing boys not only by abstaining from intercourse before ovulation but by shying away from all tight-fitting, well-insulated underwear. The ice is optional. (Chapter Six.)

Q: Can the sort of occupation or environment that a person works in have any effect on the sex of his offspring?

A: There is only sketchy information available on this. Physique, emotional factors, diet, and environment could all have some influence on acidity and alkalinity within the body and this, in turn, could conceivably have some effect on sex of offspring. Deep-sea divers are reported to produce an unusually high number of girls; individuals living at high altitudes tend to be infertile and, probably as a result of low sperm counts, produce more girls than boys. Heat, as we have just seen, can also result in low sperm count, because it raises the temperature within the scrotum. (Chapters Six and Seven.)

Q: I use an IUD for birth control. Can I test my menstrual cycle with it in place or does it change the nature of the cycle?

A: We know of no data to suggest that your cycle will significantly be different with the IUD in place. But just to be on the safe side, wait a couple months after it is removed before you try to conceive.

CHAPTER 9

Developing the "Perfect" Method: Segregation or Abortion?

For decades scientists around the world have been determined to devise a foolproof, fail-safe method of sex selection, one that would guarantee you the boy or girl you want and require very little effort on your part. Progress is being made in that direction. Just in the past few years, research in this domain has made so many strides that we would be naive to assert that the final answer won't show up within the next five or ten years.

Most of the new research efforts begin with observations about the difference in size and behavior between the male-producing and female-producing sperm. For example, the Soviets hypothesized that the two types of sperm might have different electrical properties (all living cells have an electrical potential or "charge") and that possibly they could be separated rather like protons and electrons. Dr. Manuel Gordon of Michigan State University, who followed up on this intriguing idea, put rabbit sperm in a mild saline solution and then passed a weak electrical current through it. The male-producing sperm tended to migrate toward the negative pole, while the female type generally headed for the positive pole. Inseminating rabbits with cells from the negative pole, Dr. Gordon produced males 64 percent of the time; using the sperm that migrated to the positive pole, he got females 71 percent of the time. There have been no conclusive experiments using human sperm as yet, but "electrophoresis," as this technique is called, may yet become a useful method for separating the two types of sperm.

Others have tried centrifugation techniques that exploit the difference in mass between the two types of sperm. The centrifuge spins the sperm cells in seminal solution until the two types become stratified and fall into a particular zone according to mass. Swedish scientists were the first to have any success with this, achieving the birth of eleven male calves in a row by using sperm of lesser mass. But other researchers have not been able to repeat these experiments with such success.

We discussed sedimentation techniques earlier in this book. With that method the sperm solution is allowed to "settle" under certain controlled environmental conditions, with the result that the heavier female-producing sperm sink to the bottom and the lighter male-producing sperm tend to remain suspended in the upper strata. Again, while some significant successes have been achieved in animal experiments, this method has yet to yield results that are competitive with the Shettles technique. And, of course, in order to be practical for humans, all of these procedures require artificial insemination.

Medical World News recently reported:

> Fertility clinics at the University of California Hospital in Berkeley and Michael Reese Hospital in Chicago have begun inseminating women with semen from their husbands that has undergone special processing to enhance its titer [concentration] of highly viable, highly motile spermatozoa. The method—patented and controversial—consists essentially of filtering semen through a column gradient of bovine serum albumin and skimming the cream of the hardiest and fastest swimmers. A by-product of this sperm-isolation procedure, claims its inventor, reproductive physiologist Ronald J. Ericsson, is that the viable fraction consists predominantly of Y-body sperm cells. Thus, a fallout of the fertilization trials, if they are successful, will be an inordinately high proportion of boy babies born.

We discussed Dr. Ericsson's research earlier, showing that many of his findings coincide with those of Dr. Shettles. And we would be surprised if the method outlined in *Medical World News* does not yield the excess of male births predicted. As Dr. Shettles has shown, the faster swimmers are usually the male-producing androsperm. This technique, however, is presently being used only in the treatment of infertile couples at the two centers named. Dr. Shettles' own research with very similar procedures shows that the results, in terms of sex selection, are not likely to be any better than those obtained with his present method. Moreover, the technique devised by Dr. Ericsson does nothing to help those interested in *female* offspring. Still, we believe Dr. Ericsson's research is promising and innovative.

In the meantime, Dr. Shettles, working under an Educational Foundation of America grant, is taking several
different tacks in pursuit of the "perfect" sex-selection
method. He is simultaneously investigating means of separating sperm according to differences in their density (by
centrifugation and by suspension in what are called "differential density gradient tubes"); by differences in their
migration velocities through various solutions (such as
cervical mucus, ovarian cyst fluid, and specially prepared
laboratory solutions); by differences in their head configurations (using stainless steel filters of varying sizes to
separate the two sperm populations); by differences in their
resistance to such factors as heat, ultrasound, and acidity;
and by differences in their rates of production in varying
circumstances.

All of these research efforts are yielding new possibilities.
Dr. Shettles has found that certain ultrasonic intensities
totally incapacitate the female-producing sperm while leav-
ing the male-producing population largely unaffected.
And his recent filtration experiments have been particularly rewarding, as shown by some of the results recently
published in *The Journal of Urology*. Dr. Shettles devised
a technique whereby sperm of the proper size could pass
through filters of varying configurations into ovulatory
cervical mucus. The objective was to devise a filter that
would hold back the female-producing sperm but give
passage to the male-producing type. This is not as easy as
it may sound, for one is dealing with very tiny objects, and
the difference in size between male-producing and female-

producing sperm is not great and can vary somewhat from one individual to the next.

As reported in *The Journal of Urology,* however, Dr. Shettles enjoyed a marked degree of success. In this series of experiments, twenty-eight sperm samples were exposed to the filtering system. After an hour, the fluid beyond the filter was stained with fluorochrome quinacrine and viewed under fluorescence microscopy. The quinacrine does not affect the female-producing sperm but makes the Y chromosome of the male type shine or fluoresce. In the twenty-eight samples thus tested, *no fewer than 89 percent of the human sperm that passed through Dr. Shettles' "segregator" were of the male-producing variety.* In many of the samples the count was significantly higher than that—*ranging up to 97 percent.*

Dr. Shettles concluded, in *The Journal of Urology:* "The interstices of the maze of passages through the millipore filter apparently permit the more round and smaller-headed Y-bearing sperm to maneuver the cannules far more easily and readily than the larger, more elongate-headed X-bearing spermatozoa." He found, moreover, that by all indications the filtered sperm "were actively motile and appeared cytologically intact."

Sperm thus separated could be used in artificial insemination to produce male offspring nearly 100 percent of the time, Dr. Shettles believes. Eventually he hopes to devise techniques that will enable him to artificially inseminate patients with predominantly female-producing sperm, as well.

Dr. E. James Leiberman of the National Institutes of Health once suggested that women will some day have at their disposal "a special diaphragm that will let through only the sperm that carries, let's say, the male sex and hold back those that carry the female sex." Selecting sex, then, would become merely a matter of which diaphragm a woman chooses to wear. When Dr. Leiberman made this suggestion only a few years ago, it sounded like science fiction. Now, with Dr. Shettles' filtration successes, such a development seems quite likely.

The final answer, however, is probably the one foreseen by Dr. Charles Birch, head of the Sydney (Australia) University School of Biological Sciences. He predicts that science will one day come up with a pill to determine sex. If male offspring are desired, Dr. Birch says, the husband will take one of the "little boy pills," just before intercourse or, if a female is wanted, a "little girl pill." When you consider the effectiveness of such commonplace chemicals as vinegar, baking soda, caffeine, and potassium iodide in sex selection, Dr. Birch's prediction doesn't sound so utopian or out of reach, after all.

Unfortunately it is possible today to make *certain beyond a doubt* that the next baby you give birth to will be of the sex you desire. We say "unfortunately" because we believe the technique uses abortion in an unjustifiable way. In a procedure called amniocentesis, a needle is inserted through the pregnant mother's abdominal wall to draw off some of the fluid surrounding her developing

fetus. This fluid contains fetal cells which, under microscopic examination, will reveal whether the baby is a boy
or a girl. If the fetus is not of the desired sex, the parents
could opt for abortion and try again—until they succeed
in getting a child of the sex desired, as confirmed by
amniocentesis.

It has been assumed that not many parents or their
doctors would use amniocentesis for this purpose, especially since the technique cannot pick up the required fetal
cells until relatively late in pregnancy—sometime in the
second trimester. But now there is a newer technique that
would enable doctors to determine the sex of the baby during the earliest weeks of pregnancy, when abortion is more
simply and routinely performed. This may encourage
some, perhaps many, couples to use abortion as a means of
controlling the sex of their offspring.

Consider this 1975 report from China in *Medical World
News:*

> Of 100 pregnant women recently screened to ascertain
> the sex of their unborn children, a female fetus was de-
> tected in 46. Twenty-nine of these mothers elected to
> abort. Of 53 found to be carrying males, only one woman
> chose to terminate her pregnancy. In one case, it was not
> possible to predict fetal gender. This experimental sex
> prediction service relied not on amniocentesis but on first-
> trimester non-invasive sampling of chorionic villi sloughed
> in the cervical canal. . . . Accuracy of gender prediction
> reached 93.9 percent, with the margin of error ascribed
> by the investigators to their inexperience rather than to

the cervical smear technique itself. . . . This sex-prediction program was started "to help women who desire family planning," according to the report from the Department of Obstetrics and Gynecology at the Tietung Hospital of Anshan Iron and Steel Company, People's Republic of China (*Chinese Medical Journal,* March, 1975).

The use of this new medical technology in emergency situations or to help limit population, as in China (where because of the strong cultural desire for sons a couple will continue to have children until they have at least one boy), may be justifiable and even commendable. But for routine sex selection in countries like our own, we cannot endorse its use, for reasons elaborated upon in Chapter Ten. But lest we be accused of "sour grapes" over someone else having come up with a better means of giving people children of the sex they want, we hasten to point out that the "someone" who discovered this method was Dr. Shettles himself.

In 1971, Dr. Shettles published a paper in *Nature* that described, for the first time anywhere in the world, a method by which the sex of the fetus could be determined even very early in pregnancy by obtaining cells that reveal the sex of the unborn baby. Unlike amniocentesis, this technique does not require a risky invasion of the womb. Dr. Shettles' predictions of fetal sex, based on his analysis of these cells, were 100 percent accurate.

It wasn't until 1975 that other researchers, in China and at Indiana University's Department of Medical Genetics, were able to duplicate Dr. Shettles' success in this regard.

The Indiana researchers noted the significance of his findings, not for the purposes of selecting sex via abortion (which had never been Dr. Shettles' intention), but for telling genetic counselors the baby's sex in instances where the parents might pass on a genetic defect to one sex but not to the other. (Thus abortion would come into play again, but for a more justifiable reason, if the cell test revealed that the fetus was of the vulnerable sex and would probably be severely defective.) Many sex-linked defects have been identified, but until now there has been no *easy* means of detecting the sex of unborn babies that are likely to be at risk.

In our final chapter we discuss in more detail some of the sociological and "bioethical" issues raised by our growing ability to preselect the sex of our children.

CHAPTER 10

But Is It Moral? The "Bioethical" Debate

Post-conception methods of sex selection, we believe, are morally unjustifiable. These are the methods discussed in the last chapter whereby the sex of the unborn child is determined by sampling and studying the amniotic fluid and other material, after which an abortion is performed if the fetus is of the "wrong" sex.

The scarce technical skills and equipment required to carry out these tests should be put to use where they are badly needed—in genetic counseling, where couples known to be carriers of specific sex-linked genetic defects are at risk of passing those defects on to the next generation *if* they give birth to a child of the vulnerable sex, which may be male or female, depending upon the particular defect. But beyond that we feel that to use abortion for sex selection is to commit a callous and inhuman act, particularly when a *pre-conception* method of sex selection is available that offers excellent chances of success.

There are other considerations, too. Amniocentesis is *not* totally risk-free, either to mother or fetus. And since amniocentesis cannot be used to determine sex until the second trimester of pregnancy, an abortion will be that much more perilous. Recent New York State mortality statistics reveal that second-trimester abortions are six times as life-threatening as first-trimester abortions.

In short, we wholeheartedly agree with the statement issued by the world conference on Prenatal Diagnosis of Genetic Disorders of the Fetus (conducted in Stockholm in 1975): "The group in general is not sympathetic to prenatal sex determination with the objective of aborting the fetus if it is not of the desired sex. . . . Amniocentesis is never justified merely to satisfy curiosity." To use amniocentesis and abortion to effect routine sex selection would be, in the words of Dr. Kenneth L. Vaux, professor of ethics and theology at Baylor College of Medicine, "a ghastly misuse of technology." We agree.

But we most certainly *do not agree* with those who would condemn and forbid any and all means of sex selection. The abortion issue, which forms the crux of our opposition to post-conception means of selecting sex, plays no role in pre-conception methods of the sort we advocate. Here we side with Dr. Margaret Mead, the famed anthropologist. Dr. Mead has strongly advocated sex selection by means other than abortion. Given widespread availability of sex-selective methods, she concludes, "For the first time in human history, girls would be as wanted as boys." Or to modify slightly the slogan of Planned Parenthood: "Every baby a *wanted* baby—of the desired sex."

Nor can we agree with those who believe that to choose

the sex of our offspring is wrong because it interferes with Nature or "God's will." Here we are in close rapport with the philosophy of Dr. Joseph Fletcher of the University of Virginia School of Medicine's Department of Ethics. In the following quote Dr. Fletcher is addressing himself specifically to the issue of "test-tube babies," but his reasoning applies equally to sex selection. He states:

> It seems to me that laboratory reproduction is radically human compared to conception by ordinary intercourse. It is willed, chosen and controlled, and surely these are among the traits that distinguished *homo sapiens* from others in the animal genus. . . . With our separation of baby-making from love-making, both become more human because they are matters of choice, not chance. This is, of course, essentially the case for planned parenthood. I cannot see how either humanity or morality are served by genetic roulette.

There are other objections to sex selection that we must take issue with. The prevailing fear has been that this ability to choose sex will result in a terrible imbalance between the sexes, resulting in men being more numerous and dominant than ever before—with all sorts of dire ramifications, ranging from more crime to the downfall of the Republican Party! We believe we can show these imaginative fears to be groundless.

But before we present our further defense of sex selection, let's give the other side its say. It *is* true, after all, that history records a persistent desire for sons throughout most of the world. The Romans, for example, had to pay a fourth of an *as* in taxes for every girl child born to them, but only a sixth of an *as* for every boy. Moses ruled that a woman is "unclean" for a full fourteen days following the birth of a girl but for only a week following the birth of a boy. The Jewish Talmud declares that "when a girl is born, the walls are crying," and the *Holy Book of Islam* observes that "when an Arab hears that a daughter has been born to him, his face becomes saddened." Among the Conibos of South America, a husband traditionally responds to the birth of a girl by spitting on his wife's bed. Even today, in many kingdoms, the birth of a prince is announced with two or three more cannon blasts than a princess gets.

All of this has led some observers to believe that our new ability to choose the sex of our children will result in a bumper crop of boys—with various unpleasant consequences. One man who is concerned about some of the possible results of sex selection is Dr. Amitai Etzioni, a professor of sociology at Columbia University. In an article published in a scientific journal ("Sex Control, Science, and Society," *Science* 161, Sept. 13, 1968), he addressed himself to the probability that a great many parents will be taking advantage of sex-selection methods "five years from now or sooner." Then he cited studies showing that the demand for male children is 55 to 65 percent greater than for females. In light of this, he fears "an overproduction of boys" and predicts that this "will very likely affect most aspects of social life." He goes so far as to say that parental control over sex of offspring could bring to an end the two-party political system and throw the country back into a frontier atmosphere.

This may not be as farfetched as it sounds, *provided* Dr. Etzioni's basic premise—that parents will consistently overproduce males, given the chance—is correct. But before returning to that issue, let's see what else Dr. Etzioni has to say: he notes that men vote "systematically and significantly more Democratic than women." Since the Republican Party has been steadily losing support over the last generation, according to the professor, he believes that "another five-point loss could undermine the two-party system to a point where Democratic control would be uninterrupted."

First of all, it is safe to say that Dr. Etzioni does not believe that sex selection is a Democratic Party plot. His theory is certainly an intriguing one, but it seems difficult to believe that if the Republican Party should fall by the way it would be simply because of an overproduction of males. No one can say for sure what the political orientation of future male generations will be. Then, too, many of Dr. Etzioni's colleagues have said that women are becoming more and more the dominant sex, while men are receding more and more into the background and even taking up household duties. Perhaps in the future, the activist woman will associate almost exclusively with the labor-oriented Democratic Party, while the domesticated male will take up residence in the Republican Party. Or suppose that the two-party system should temporarily col-

lapse as a result of heightened male supremacy. Men have
about as many disagreements as women and can be counted
on to go their separate ways—via new political parties—at
some point in the future. Finally, many studies indicate
that party ties are coming to mean less and less to voters;
issues and personalities play an ever greater role in politics,
which may be one reason that we have recently had Republican presidents despite the great numerical superi-
ority of registered Democrats.

Dr. Etzioni goes on to point out that women as church-
goers, consumers of "culture," and so on, do more than
men to maintain what is generally known as "civilization."
Hence he concludes that "a significant and cumulative
male surplus will . . . produce a society with some of the
rougher features of a frontier town." This new frontier,
he predicts, will be populated by a lot of single men on
the make (sort of latter-day saddle bums, gunslingers, and
cowpokes) for the few available females. To take up the
slack, he goes on, prostitution and homosexuality will increase substantially. Further, he foresees an increase in in-
terracial and interclass tensions. Minority groups and the
lower classes put a higher premium on male offspring than
do the more affluent, and these groups in particular, Dr.
Etzioni believes, could be counted on to use sex-selection
procedures. This bumper crop of lower-class boys, he says,
would then be forced to seek out girls from higher status
groups, thus increasing class and race strife.

There is one major chink in Dr. Etzioni's reasoning.
The lower classes, who tend to be less literate, more suspi-

cious and fearful of anything that "interferes" with basic biological functions, and more influenced by superstition and folk medicine, will unfortunately be the last to take advantage of sex selection on any significant scale. Birth-control pills, for example, could be of greatest benefit to the poor. Yet it is the more affluent and the more sophisticated who are the greatest consumers of birth-control devices and chemicals.

Commentators other than Dr. Etzioni have added other fears to the list, fears once again based on the notion that sex-selection techniques will help fuel boy-girl fads that will have "disastrous" consequences. If female offspring become the rage, they say, society will have to give its blessings to polygamy; if boys become the vogue, then polyandry would have to prevail. "The dangers are not apocalyptic," Dr. Etzioni concedes, "but are they worth the gains to be made?"

Hopefully, the preceding paragraphs have demonstrated that the "dangers" have been somewhat exaggerated, even if one accepts Dr. Etzioni's basic premise—that parents will overproduce one sex. If that premise is untrue, of course, then the alleged dangers completely dissolve.

Dr. Shettles is convinced, from personal experience, that parents will *not* use his techniques to produce only sons— or only daughters. "Over the years, parents have expressed only one desire," he says, "and that is to have families that are well balanced in terms of sex. Most find an equal number of boys and girls ideal." The same attitude emerges from the hundreds of letters that arrived following early publicity of Dr. Shettles' work: families with three boys, for example, wanted nothing so much as a girl, while those with a preponderance of girls were equally enthusiastic about being able to beget boys.

Many couples said they had initially planned for a family of two children, hoping for one of each. But when both offspring turned out to be of the same sex they made a third attempt, and so on. "We now have four boys," one woman wrote, "and that is two more than we really wanted. If we had known of this research several years ago we would never have ended up in this situation. Of course, we love our sons and have only ourselves to blame. But we just kept thinking: once more and this time it will have to be a girl."

So it is not too farfetched to envision sex selection making a significant contribution in the effort to control the population explosion. How much better it would be to achieve the ideal family balance in two tries instead of three or four or more, or never.

In any event, to return for a moment to Dr. Etzioni, it seems that overproduction of one sex, if that should occur, would ultimately create a reaction, causing parents to start choosing the other sex again. As soon as something useful, whether it be beefsteak or females, is in short supply, there is almost always a huge demand for it, followed by a massive effort to produce the desired commodity in volume sufficient to meet that demand.

G. Rattray Taylor, in his book *The Biological Time-Bomb*, takes an optimistic point of view about this: "It may be," he writes, "that in Western societies there would be some slight preference for a son, expressed perhaps as a tendency on the part of some people to have sons only, more often than daughters only. I suspect that this tendency would not be so marked that it could not be checked by propaganda and good sense."

Dr. Shettles' conviction that sex selection will not result in troublesome imbalances of the sexes has been strongly supported by the most definitive study yet conducted on the issue. Dr. Charles F. Westoff of the Office of Population Research at Princeton University and Dr. Donald R. Rindfuss of the Center for Demography and Ecology at the University of Wisconsin surveyed nearly 6,000 married women with varying numbers of children. They found that if sex selection were routinely used today there would be an excess of male births during the first two years (with a backlog of childless women in the process of using preselection to have boys as their first child), "followed," as *The New York Times* put it, "by a wave of female births to achieve balance." These "oscillations would eventually disappear, so that the sex ratio would be similar to the present natural sex ratio," concluded the *Times* in its report of the Westoff/Rindfuss study.

There is no denying that a majority of the couples want sons *first*—but some sociologists have clearly erred in assuming that sons are *all* they want. Other studies have concurred with the Westoff/Rindfuss study. Dr. Otfried Hatzold of Munich, Germany, recently surveyed more than 1,000 couples and found that of the 365 who had one son and no daughters, 352 wanted girls to complete their families. Of 104 couples who had two or more sons and no daughters, all 104 said that if they had more children they would like to have daughters. Similarly, of 468 who had one daughter and no sons, 467 said they hoped their next child would be a boy. Of 133 couples with two or more daughters and no sons, only one wanted another daughter.

The "dangers" of sex selection simply have not been demonstrated. The *advantages,* on the other hand, are manifest: parental satisfaction, smaller, more balanced families, and even *healthier* families.

Where does health come in? Sometimes health—or lack of it—is attached to our sex chromosomes. Only males, for example, suffer from hemophilia, the grim and often fatal "bleeder's disease" that is not so rare as many people think. Similar hereditary, "sex-linked" diseases include one type of muscular dystrophy and numerous enzyme-deficiency disorders that can kill, cripple, and retard for life.

Though most of these diseases remain incurable, they can be prevented by sex selection. Take hemophilia, for example, often called the "disease of kings," because it afflicted so many men in reigning European families. This disease leaves its victims very nearly defenseless against even small cuts and wounds, because their blood lacks the factor that enables it to clot effectively. The recessive gene that determines whether a person will suffer from this disease is carried, when it exists at all, in one of the two X chromosomes of each immature female sex cell. After final cell division, only half of these cells carry the recessive gene. If one of these carrier eggs is mated with an X-bearing (female-producing) sperm, the female offspring will not suffer from the disease, because the normal gene, inherited from the father, is dominant. If, however, the carrier egg is fertilized by a Y-bearing sperm, the disease will manifest itself because the Y chromosome does not carry the gene. Doctors can now tell women who are known carriers of hemophilia that their offspring, if male, will have a 50-50 chance of suffering from the disease.

Many women who discover that they are carriers of sex-linked diseases such as hemophilia decide to have an abortion when they learn that they are going to give birth to male offspring. Such abortions could be avoided altogether if these carriers of sex-linked diseases could simply avoid conceiving children of the vulnerable sex.

The value of sex selection in helping to overcome these diseases motivates many researchers in this field, such as Drs. Robert Edwards and Richard Gardner of Cambridge University. Writing in *New Scientist,* they point out that "the elimination of these disorders in one generation, by a judicious choice of the sex of the offspring, would not only be of direct benefit to that generation, but would benefit the race for generations to come."

What other advantages might accrue from our ability to choose the sex of our children? One medical man, Dr. A. L. Benedict, has suggested that sex selection might have some psychological benefits beyond the obvious ones that come from having sexually balanced families. Some parents, Dr. Benedict believes, are really suited to raising chil-

dren of only one sex. It may be that a woman who has a strong aversion to little girls is mentally disturbed and shouldn't really have *any* children. But since she is likely to go ahead no matter what we think, isn't it better if she has nothing but boys—better for her children as well as herself?

Similarly, there are men who have such a strong desire to beget children of one sex that offspring of the "wrong" gender suffer for it. The father's disappointment is quickly communicated to the child, very often with no words ever spoken; as many a psychiatric case history has shown, the child may feel at once rejected and guilty—for having "failed" to be born a member of the opposite sex. The child may then either withdraw into himself or perhaps try to correct his error by acting as though he were indeed a member of the opposite sex. If the situation continues to deteriorate in this way, the child may be psychologically scarred for life, unable to function properly in his biologically assigned sexual role.

Beyond this, sex selection could come in handy in a number of situations in which, for some reason, there is a particular shortage of one sex or the other. In the days of the woman-scarce Old West, for example, there was probably a big demand for females, at least in the towns where single men congregated on weekends. On the frontier farms, of course, boys would probably have prevailed, since they could help with the heavy work. If we move out into the new frontier of space it is possible that certain circumstances and environments will again call for the proliferation of one sex or the other—and this time the means to meet the demand will be at hand. We will be similarly prepared if war or disease should destroy a great many men or women. Perhaps in the future, governments will call upon their constituents to fulfill their patriotic duties, not so much by paying their taxes as by producing either boys or girls, whichever happens to be needed as a result of some catastrophe.

For the moment we can content ourselves with the fact that a method of sex selection has been developed and is at our disposal. Now parents have the opportunity to make a scientific attempt at choosing the sex of their children and to make that attempt with a justifiably high expectation of success. The procedures involved are safe and simple, and nothing about them is morally or ethically objectionable. Protestant ministers have inquired about the procedures with the intention of incorporating them into their own family planning, rabbis have cooperated with Dr. Shettles in his research, and the Roman Catholic Church has bestowed its blessing. Monsignor Hugh Curran, director of the family-life bureau of the Archdiocese of New York, says that the Church has no objections to Dr. Shettles' sex-selection procedures "as long as the intent of these efforts is not to prevent conception."

We are at a pivotal point in our evolutionary development. For the first time, what we call "the facts of life" need no longer govern us. We possess the knowledge to alter those "facts," and Dr. Fletcher of the University of

Virginia is one who argues that we have an ethical mandate to do so.

"We cannot accept the 'invisible hand' of blind, natural chance or random nature in genetics," he declares, "any more than we could [accept] old Professor Jevons' theory of feast and famine in nineteenth-century *laissez-faire* economics, based on sun spots and tidal movements. To be men we must be in control. That is the first and last ethical word. For when there is no choice, there is no possibility of ethical action. Whatever we are compelled to do is amoral."

Acid/Alkaline Foods

Dr. R. C. Pilsner, an advocate of a nutritional approach to health, has written that "rest and sleep are alkalizers. So is exercise, fresh air, pleasure, laughter, conversation, enjoyment —even love! Acidifiers are worry, fear, anger, gossip, hatred, envy, 'crabbing,' selfishness, and silence—also love-hunger." We agree. As we pointed out earlier, women who are under continual stress are likely to have a more acidic vaginal milieu, an environment that is hostile to sperm penetration and particularly to male-producing androsperm penetration. And men who are under various forms of stress are more likely to have lower sperm counts. Extreme stress, both physiological and psychological, can result in various forms of male and female infertility.

The food you eat can be stressful, too, if you are getting a poor balance of acid and alkaline foods. Inasmuch as such substances as caffeine, potassium iodide, baking soda, and vinegar can have an impact on sex ratios, we believe that diet should not be entirely overlooked. (Consider, too, the reports we cite showing that women exposed to LSD almost always give birth to females.) What we take into our bodies *can* have an effect on the sex of our offspring. Our knowledge in this area is rudimentary (the nutritional sciences in general are still very young), but it appears that a "safe" diet is one in which a *majority* of your foods are alkaline. This is true for *both* men *and* women whether you are interested in conceiving boys or girls. Never let the acids in your diet predominate, even if you are trying for a girl. If you are trying for a boy, however, you may safely and beneficially increase your relative intake of alkaline foods, if you like.

The following are commonly listed as alkaline or acidic. Please note that some fruits, such as limes, are generally regarded as acidic but in the body actually have an alkaline effect.

Alkaline Vegetables

Beans (string, lima, green, sprouts)
Beets
Broccoli
Cabbage
Carrots
Cauliflower
Celery
Chard
Collards
Cucumber
Eggplant
Endive
Garlic
Leek
Lettuce
Onions
Parsely
Parsnips
Peppers (green and red)
Potatoes
Pumpkin
Radish
Spinach
Squash
Turnips

Acidic Vegetables

Artichokes
Asparagus tips
Beans (dried)
Brussel sprouts
Garbanzos (chick-peas)
Lentils
Rhubarb

Alkaline Fruits

Apples
Apricots
Avocados
Bananas (ripe)
Berries (fresh)
Cantaloupe
Cherries
Currants
Dates
Figs
Grapes
Grapefruit
Guavas
Lemons
Limes
Melons
Nectarines
Oranges
Papayas
Peaches
Pears
Pineapple (ripe)
Raisins
Tangerines
Tomatoes

Acidic Fruits

Bananas (green)
Cranberries
Olives (green, pickled)
Preserves, jellies, and canned, sugared, dried, sulphured, or glazed fruits
Plums (slightly acidic)
Prunes (slightly acidic)

Alkaline Dairy Products

Buttermilk
Raw milk (human, cow, goat)
Whey
Yogurt

Acidic Dairy Products

Butter
Cheese
Cottage cheese
Ice cream
Milk (other than raw)

Alkaline Meats

None

Acidic Meats

All meat, fish, fowl

Alkaline Cereals	**Acidic Cereals**
Corn (on the cob)	Buckwheat
	Barley
	Breads (all flours)
	Corn (other than on the cob)
	Macaroni and spaghetti
	Noodles
	Oatmeal
	Rice

Alkaline Nuts	**Acidic Nuts**
Almonds	All other nuts
Chestnuts (roasted)	
Coconut (fresh)	

Other Alkaline Foods	**Other Acidic Foods**
Alfalfa products	Alcoholic beverages
Coffee substitutes	Candies
Honey	Cocoa and chocolate
Kelp	Coca-Cola, other soft drinks
Tea (unsweetened)	Dressings
Yeast cakes	Egg whites
	Spices
	Tobacco
	Vinegar

Some authorities persuasively argue that the American diet has become much too acidic, with overemphasis on meat (all of which is acidic), sweets, pastries, starches, and preserved foods. Establishing a diet made up of 70 to 80 percent alkaline foods, these authorities argue, should result in greater nutritional health and greater resistance to a variety of ills. This is one of the arguments contributing to the trend toward a modified vegetarian diet.

Reader Questionnaire

This questionnaire should be filled out as soon as you know that you are pregnant. Save the questionnaire until the birth of your child so that we may know the result of your sex-selection effort, too. Mail to David M. Rorvik, Dodd, Mead & Company, Inc., 79 Madison Avenue, New York, N.Y. 10016

Date _____

Wife's name_____ Age _____ Occupation _____

Husband's name_____ Age _____ Occupation _____

Number of children prior to this attempt _____

Sex of those children _____

Was this your first attempt at sex selection? _____

If "no," did you succeed or fail on other attempts? (Please enu-
        merate)_____

_____

In this attempt, what sex did you desire? _____

Did you succeed?_____ If "yes," give name of child, date of
   birth, weight at birth _____

   _____

Was your pregnancy or delivery complicated in any way, even
   though you did succeed? If "yes," please specify _____

   _____

   _____

If you did *not* succeed, did you suffer a miscarriage or other com-
   plication that terminated pregnancy? (Please specify nature of
   difficulty) _____

   _____

If you delivered a live baby and the child was not of the sex you
   desired, do you believe you carefully carried out all the recommendations in this book? _____

What form(s) of birth control did you use in the twelve months
   prior to becoming pregnant in this attempt? (Break this down as
   "first seven months the pill; after that condoms") _____

   _____

For how many menstrual cycles did you keep basal body temperature charts prior to making your effort? _____

Provide the following information about each menstrual cycle you

charted:

	Total length/days	Number of bleeding days	Day of upward temperature shift
Cycle 1			
Cycle 2			
Cycle 3			
Cycle 4			
Cycle 5			
Cycle 6			

(If you tested for more than six cycles, include additional infor-

mation on another sheet of paper. Attach actual charts for

all cycles, if possible.)

Did you use Tes-Tape to assist in finding your ovulation time? _____

Was it generally useful and easy to interpret? _____

Do you experience *Mittelschmerz?* _____ If so, did you make

use of it in finding your ovulation time? _____

Did you use the "stretch test," i.e., did you observe changes in the

cervical mucus to assist in finding ovulation time? _____

Was this helpful to you? _____

Which was most helpful—the temperature charting, the Tes-Tape,

Mittelschmerz, or the stretch test? _____

Were you confident that you had found your ovulation date when
you made the attempt that resulted in pregnancy? _____

If "no," explain _____

_____

If you have a chart (or can recreate a chart) showing the cycle in
which you became pregnant in this attempt, *please attach it and
send it to us.* (This is very important.) On it, indicate the day
you believe you ovulated. Also indicate the day you had intercourse (if trying for a boy) or the last day on which you had
intercourse prior to ovulation (if you were trying for a girl).
Indicate here whether you have attached this chart _____

Do you (wife) have an ulcer condition or "acid stomach"?_____

Has either you or your husband suffered any illnesses, other than
colds, in the year preceding your attempt? _____

If "yes," please specify _____

Does your husband's job expose him to unusual stress such as heat,
high altitude, underwater pressures, noxious gases, or other
chemicals? _____ If "yes," please specify_____

_____

Has either of you had infertility problems in the past? _____

If so, explain _____

Has either of you ever used LSD, mescalin, or similarly potent "mind-altering" drugs? (answers confidential) _____

If "yes," which of you, when, and how often? _____

_____

Has either of you ever been diagnosed as schizophrenic? _____

Did you find any part of the procedure for sex selection distasteful, annoying, or distracting from sexual pleasure? _____

If "yes," please explain _____

_____

Was your doctor aware that you were using these methods? _____

If "yes," was he approving? _____ Disapproving? _____

Skeptical? _____ Helpful? _____ Indifferent? _____

Had you sought out medical help for selecting sex of your offspring in the past? _____ With what results? _____

_____

Do you now regard your family as complete? _____

If you had not succeeded in conceiving a child of the sex you desired this time, would you have kept trying until you did? In other words, how many more times would you have tried before giving up? _____

If you do *not* now regard your family as complete, how many more
children do you desire and of which sex? _____

_____

Will you try this method again, even if you failed with it this time?

_____

Do you believe there is anything "immoral" or otherwise wrong
about sex selection? _____ If "yes," please explain

_____

_____

Do you think sex selection will contribute to population *control?*

_____

Will it contribute to population *imbalance,* far more boys than
girls, for example? _____

How did you hear about this book? _____

Did you find the instructions sufficiently clear? _____ If "no,"
what was confusing? _____

_____

Is there anything that is not in this book that you believe should be
included in future books? _____ If "yes," what?_____

_____

Additional comments, if any: _____

_____

_____

Below are two sets of questions—one for those who wanted boys, one for those who wanted girls.

Questions to be answered only by those who were trying for a boy:
Did you and your husband abstain from intercourse from the beginning of your cycle until the day of suspected ovulation? _____ If "no," explain _____

_____

Did your husband avoid tight-fitting pants, jockey shorts, and jock-straps prior to the attempt? _____

Did he drink coffee prior to the attempt? _____

Did you take potassium iodide? _____

Did intercourse take place the morning after the temperature had shifted upward? _____ Or did it take place on the day when the last low temperature before the upward shift was recorded? _____

At what time of day did this intercourse take place? _____

For how long *after* the suspected day of ovulation did you abstain from intercourse? _____

When you did resume sexual relations, did you use any form of contraception? _____ If "yes," what kind? _____

Did you (the wife) have orgasm? _____ If "yes," did this pre-cede your husband's orgasm? _____ Coincide with it? _____

Follow it? _____

Was penetration from the rear? _____

Was deep penetration at the time of male orgasm achieved? _____

Is there, on either side of the family, a predominance of one sex

among offspring? _____ If so, explain _____

_____

Did you both want a boy with equal intensity, as far as you can

tell? _____ If "no," which of you wanted a boy more? _____

Why did you want a boy? _____

_____

Questions to be answered only by those who were trying for a girl:

How many times did you have intercourse from the first day of

your cycle up to the cut-off date prior to ovulation? _____

Did you use contraception on any of those days?_____ If

"yes," why did you use contraception? _____

_____

At what time of day did you last have intercourse on the cut-off

date? _____

Did you schedule the cut-off date a certain number of days from the

probable time at which the *last low* temperature of the cycle

would be recorded or from the day upon which the *upward*

temperature shift would probably be noted? _____

Did you initially schedule the cut-off date *three days* before suspected ovulation? _____

Did you succeed in becoming pregnant on this schedule? _____

    If "yes," how many attempts (cycles) did it take before you became pregnant? (Do not count practice cycles in which no serious effort was made to become pregnant) _____

If you did not succeed in becoming pregnant with a three-day cut-off, how many times did you try before giving up and going to a shorter cut-off schedule? _____

If you didn't succeed at three days before ovulation, at what interval did you make the next attempt: two and one-half days? two days? _____

Did you succeed using *this* cut-off interval? _____ If "yes," how many cycles did it take before you became pregnant? _____

    If "no," at what interval *did* you finally become pregnant? (whether with a boy or girl) _____

How long did you abstain from intercourse *after* the cut-off date?

_____

When you resumed sexual relations, did you use contraception? _____ If "yes," what type did you use?_____

During the attempt, was intercourse preceded on each occasion by the prescribed acidic douche? _____ If "no," why not? ___

_____

Did you (the wife) experience orgasm during any episode of inter-

course during any attempt to achieve a girl? _____

Was the face-to-face position used during intercourse? _____

Did your husband succeed in shallow penetration at the time of his

orgasm? _____

Is there, on either side of the family, a predominance of one sex

among offspring? _____ If so, explain _____

_____

Did you both want a girl with equal intensity, as far as you can

tell? _____ If "no," which of you wanted a girl more? _____

_____

Why did you want a girl? _____

_____

_____

Thank you for your cooperation. We are confident that your answers will assist us significantly, not only in assessing our successes and failures but in devising refinements for future publications. We hope that you will consent to let us use your names and case histories in those future publications. If you do not want your names used, we will respect your wishes, but we urge you to fill out these forms, clip them from the book and mail them to us. Every case counts, whether or not there are names attached.

Bibliography

UNSIGNED REPORTS

Boy or girl, take your pick. *Science Digest,* April, 1974.
Coital patterns successful in predicting child's sex. *OBGYN News,* February 15, 1971.
Male sperms swifter off the mark. *New Scientist,* July 20, 1967.
Of coitus and the baby's sex. *Medical World News,* August 13, 1972.
"Preselecting" infant's sex may help prevent birth defects. *OBGYN News,* October 15, 1975.
Psychotomimetics affect sex ratio? *OBGYN News,* December 15, 1970.
Separation of X and Y spermatozoa. *Research in Population,* January, 1974.
Sperm shape whips up a storm. *Medical World News,* August 12, 1960.
Toward sex on order. *Time,* June 27, 1960.
Two shapes of sperm found in humans. *The New York Times,* June 5, 1960.

SIGNED REPORTS

Benendo, Franciszek. The problem of sex determination in the light of personal observations. *Polish Endocrinology* 21 (1970): 200–207.
Bennett, D. and Boyse, E. A. Sex ratio in progeny of mice inseminated with sperm treated with H-Y antiserum. *Nature* 246 (November 30, 1973): 308–309.

Bhattacharya, B. C. Sex control in mammals. *Zeitschrift für Tier-zuchtung und Zuchtungsbiologie* 72 (1958): 250–254.

Bhattacharya, B. C. The different sedimentation speeds of X and Y sperm and the question of optional sex determination. *Zeitschrift für Wissenschaftliche Zoologie* 166 (1962): 203–250.

Dahl, Roald. Ah, sweet mystery of life. *The New York Times*, September 14, 1974.

Dawson, E. R. *The causation of sex in man.* 2nd ed. London: H. K. Lewis & Co., 1917.

Edwards, R. G. and Gardner, R. L. Sexing of live rabbit blastocysts. *Nature* 214 (May 6, 1967): 576–577.

Ericsson, R. J., Langevin, C. N., and Nishino, M. Isolation of fractions rich in human Y sperm. *Nature* 246 (December 14, 1973): 421–424.

Etzioni, A. Sex control, science and society. *Science* 161 (September 13, 1968): 1107–1112.

Etzioni, A. Selecting the sex of one's children. Letter to the editor, *Lancet* 1 (May 11, 1974): 932–933.

Freedman, D. S., Freedman, R., and Whelpton, P. K. Size of family and preferences for children of each sex. *American Journal of Sociology* 66 (1960): 141–146.

Galton, Laurence. Parent and child: choosing the sex of a child. *The New York Times Magazine*, June 30, 1974.

Gordon, M. J. Control of sex ratio in rabbits by electrophoresis of spermatozoa. *Proceedings of the National Academy of Sciences* 43 (1957): 913–918.

Guerrero, R. Association of the type and time of insemination within the menstrual cycle with the human sex ratio at birth. *New England Journal of Medicine* 291 (November 14, 1974): 1056–1059.

Hart, D. and Moody, J. D. Sex ratio: experimental studies demonstrating controlled variations—preliminary report. *Annals of Surgery* 129 (May, 1949): 550–571.

Hatzold, Otfried. Personal communication, Munich, Germany. September 3, 1974.

James, W. H. Cycle day of insemination, coital rate and sex ratio. *Lancet* 1 (January 16, 1971): 112–114.

James, W. H. Sex ratios in large sibships, in the presence of twins and in Jewish sibships. *Journal of Biosocial Science* 7 (April, 1975): 165–169.

Janerich, D. T. Sex ratio and season of birth. Letter to the editor, *Lancet*, April 24, 1971.

Kaiser, R., Broer, K. H., Citoler, P., and Leister, B. Penetration of spermatozoa with Y-chromosomes in cervical mucus by an in vitro test. *Geburtshilfe und Frauenheilkunde* 34 (June, 1974): 426–430.

Keynes, R. D. The predetermination of sex. *Advancement of Science* 24 (1967): 43–46.

Kleegman, S. J. Can sex be planned by the physician? In *Fertility and Sterility*, Proceedings of the 5th World Congress on Fertility and Sterility (Stockholm, June 16–22, 1966), edited by B. Westin and N. Wiquist. Amsterdam: Excerpta Medica (International Congress Series No. 133, 1967): 1185–1195.

Kleegman, S. J. Therapeutic donor insemination. *Fertility and Sterility* 5 (1954): 7–31.

Knaack, J. Arbitrary influence on sex by sedimented bull sperms—results of a large-scale test. *Fortpflanzung Besamung und Augzucht der Haustiere* 4 (1968): 279–282.

Krzanowski, M. Dependence of primary and secondary sex-ratio on the rapidity of sedimentation of bull semen. *Journal of Reproduction and Fertility* 23 (1970): 11–20.

Lappé, M. Choosing the sex of our children. *The Hastings Center Report* 4 (February, 1974): 1.

Leff, D. N. Boy or girl: now choice not chance. *Medical World News*, December 1, 1975.

Levy, Jacob. The surplus of male births among Jews—a contribution to the question of sex-determination. *Koroth* 6 (November, 1973).

Lowe, C. R. and McKeown, T. The sex ratio of human births related to maternal age. *British Journal of Social Medicine* 4 (1950): 78–85.

Novitski, E. and Kimball, A. W. Birth order, parental ages and sex of offspring. *American Journal of Human Genetics* 10 (1958): 268–275.

Parkes, A. S. Mythology of the human sex ratio. In *Sex ratio at birth—prospects for control*, edited by C. A. Kiddy and H. D. Hafs. Champaign, Ill.: American Society of Animal Science, 1971: 38–42.

Repetto, R. Son preference and fertility behavior in developing countries. *Studies in Family Planning* 3 (April, 1972): 70–76.

Rhine, S. A., Cain, J. L., Cleary, R. E., *et al.* Prenatal sex detection with endocervical smears: successful results utilizing Y-body fluorescence. *American Journal of Obstetrics and Gynecology* 122 (May, 1975): 155–160.

Robinson, D., Rock, J., and Menkin, M. F. Control of human spermatogenesis by induced changes of intrascrotal temperature. *Journal of the American Medical Association* 204 (April 22, 1968): 80–87.

Rohde, W., Porstmann, T., and Dörner, G. Migration of Y-bearing spermatozoa in cervical mucus. *Journal of Reproduction and Fertility* 33 (April 1, 1973): 157–169.

Rohde, W., Porstmann, T., Prehn, S., and Dörner, G. Gravitational pattern of the Y-bearing human sperm in density gradient centrifugation. *Journal of Reproduction and Fertility* 42 (March, 1975): 587–591.

Schellen, A. *Artificial insemination in the human.* Amsterdam: Elsevier Publishing Co., 1957.

Schilling, E. Sedimentation as an approach to the problem of separating X and Y chromosome bearing spermatozoa. In *Sex ratio at birth—the prospects for control.* Champaign, Ill.: American Society of Animal Science, 1971: 76–84.

Schröder, V. Sex control of mammalian offspring and the biochemical and physiological properties of X and Y bearing sperm. *Animal Breeding Abstracts* 10 (December, 1942): 252.

Seguy, B. Methods of natural and voluntary selection of the sexes. *Journal de Gynécologie Obstétrique et Biologie de la Reproduction* 4 (1975): 145–149.

Shettles, L. B. Nuclear morphology of cells in human amniotic fluid in relation to sex of infant. *American Journal of Obstetrics and Gynecology* 71 (1956): 834–838.

Shettles, L. B. Biological sex differenecs with special reference to disease, resistance and longevity. *Journal of Obstetrics and Gynecology of the British Empire* 65 (1958): 288.

Shettles, L. B. Observations on human spermatozoa. *Bulletin of the Sloane Hospital for Women* 6 (1960): 48.

Shettles, L. B. Nuclear morphology of human spermatozoa. *Nature* 187 (1960): 254.

Shettles, L. B. Differences in human spermatozoa. *Fertility and Sterility* 12 (1961): 20.

Shettles, L. B. Head differences in human spermatozoa. *Journal of Urology* 85 (1961): 355.

Shettles, L. B. Human sperm populations. *International Journal of Fertility* 7 (1962): 175.

Shettles, L. B. The great preponderance of human males conceived. *American Journal of Obstetrics and Gynecology* 89 (May 1, 1964): 130–133.

Shettles, L. B. Factors influencing sex ratios. *International Journal of Gynaecology and Obstetrics* 8 (September, 1970): 643–647.

Shettles, L. B. Use of the Y chromosome in prenatal sex determination. *Nature* 230 (March 5, 1971): 52–53.

Shettles, L. B. Sperm morphology, cervical milieu, time of insemination and sex ratios. *Andrologie* 5 (1973): 227–230.

Shettles, L. B. Sex selection. Letter to the editor, *American Journal of Obstetrics and Gynecology*, January 2, 1976.

Shettles, L. B. Human spermatozoa filtration. *Journal of Urology*, in press.

Shettles, L. B. Potassium iodide enhancement of cervical and uterine secretions. *Drug Therapy*, in press.

Talwar, P. P. Effect of desired sex composition in families on the birth rate. *Journal of Biosocial Science* 7 (April, 1975): 133–139.

Teitelbaum, M. S. Factors associated with the sex ratio in human populations. In *The structure of human populations*, edited by G. A. Harrison and A. J. Boyce. London: Oxford University Press, 1972: 90–109.

Unterberger, F. Sex determination and hydrogen ion concentration. *Deutsche Medizinische Wochenschrift* 58 (May 6, 1932): 729–731.

Wachtel, S. S., Koo, G. C., Zuckerman, E. E., Hammerling, U., Scheid, M. P., and Boyse E. A. Serological cross-reactivity between H-Y (male) antigens of mouse and man. *Proceedings of the National Academy of Sciences* 71 (April, 1974): 1215–1218.

Wakim, P. E. Determining the sex of baby rabbits by ascertaining the pH of the vagina of the mother before mating. *Journal of the American Osteopatic Association* 72 (October, 1972): 173–174.

Westoff, C. F., Potter, R. B., Jr., and Sagi, P. C. *The third child: a study in the prediction of fertility.* Princeton, N.J.: Princeton University Press, 1963.

Westoff, C. F. and Rindfuss, R. R. Sex preselection in the United
States: some implications. *Science* 184 (May 10, 1974): 633–636.
Zirkle, C. The knowledge of heredity before 1900. In *Genetics and
the 20th century: essays on the progress of genetics during its
first 50 years*, edited by L. Gunn. New York: Macmillan, 1951:
35–37.

Index

Landrum B. Shettles, M.D., Ph.D., D.Sc., and Associate Professor of obstetrics and gynecology at Columbia University's College of Physicians and Surgeons and an attending physician at ,Columbia-Presbyterian Medical Center for twenty-seven years, is now Chief of Obstetrics and Gynecology at Gifford Memorial Hospital in Randolph, Vermont. Dr. Shettles, who received his M.D. and Ph.D. degrees from the Johns Hopkins University, was a Markle Scholar and has held appointments at New York Medical College, French Polyclinic Post Graduate Medical School and Health Center, and Doctor's Hospital in New York City. He has served as director of research at the New York Fertility Research Foundation. He is a Fellow of the American College of Obstetricians and Gynecologists and the American College of Surgeons. A noted researcher, some two hundred of his scientific papers have been published in leading medical journals throughout the world. He is coauthor (with Roberts Rugh, M.D.) of the widely acclaimed book *From Conception to Birth: The Drama of Life's Beginnings*.

David M. Rorvik holds a Master of Science degree from Columbia University's Graduate School of Journalism. He was the winner of a Pulitzer Traveling Fellowship to Africa in 1967 and was named an Alicia Patterson Fellow in 1976, with a grant to investigate "the politics of cancer research in the United States, Soviet Union, Japan, and China." He is a former *Time* Magazine science and medicine reporter. His science and medical articles have appeared in most major magazines. He is the author of numerous books, including (with Dr. Shettles) *Your Baby's Sex: Now You Can Choose* and (with Dr. O. S. Heyns of South Africa) *Decompression Babies*.